DAF at work:

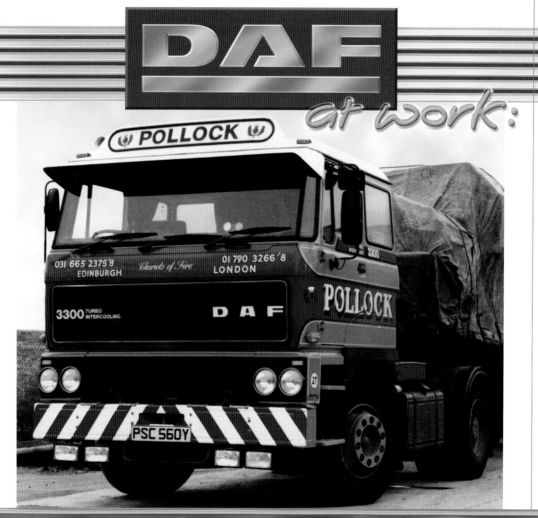

2800, 3300 & 3600

Patrick W Dyer

ACKNOWLEDGEMENTS

My grateful thanks to the following for their help and support: Tony Pain, Peter Symons, George Bennett, John W Henderson, Adrian Cypher, Marcus Lester, Dave Wakefield, Del Roll, Sue Chapman, all at Old Pond and, of course, Linda and Jess Dyer. Let's do it again soon.

ABOUT THE AUTHOR

Patrick Dyer, born in 1968, grew up during one of the most notable and exciting periods of development for heavy trucks and also the last of the real glory days for trucking as an industry. This is reflected in his subject matter. His first books covered F88 and F89 from Volvo and the LB110, 111, 140 and 141 from Scania; this new title deals with the equivalent products from DAF. Although Patrick's day job is in motor sport, he holds a Class One licence and drives whenever the opportunity arises. He is also the proud owner of a 1983 Volvo F12, which he is returning to its original condition with the help of long-term friend and noted classic vehicle restorer, Ashley Pearce.

DECLARATION

There were at least six recognised methods of measuring engine output for trucks during the period covered by this book. Manufacturers and magazines often quoted different outputs for the same engine using BS au, SAE, DIN and ISO systems, some gross and some net, much to everyone's confusion. The DAF 2800 DKTD could rightly be regarded as a 307 or 320 bhp machine depending on which literature you read, so in the interest of continuity all outputs quoted in this work are those of the manufacturer, DAF, as published at the time.

DEDICATION

This book is dedicated to all of you out there with a passion for classic trucks; let's keep their wheels turning for future generations.

First published 2010

Copyright © Patrick W. Dyer, 2010

The moral right of the author in this work has been asserted

All rights reserved. No parts of this publication may be reproduced, stored in a retrieval system, or transmitted, in any form or by any means electronic, mechanical, photocopying, recording or otherwise, without prior permission of Old Pond Publishing.

ISBN 978-1-906853-37-2

A catalogue record for this book is available from the British Library

Published by
Old Pond Publishing Ltd
Dencora Business Centre,
36 White House Road
Ipswich IP1 5LT United Kingdom

www.oldpond.com

Front cover photograph
By the early 1980s, Pollock's fleet-buying policy was controlled by three members of the family: father and founder William, and his two sons, George and Iain, each with their own favourite manufacturer. George Pollock was pro-DAF and PSC 560Y was one of six 3300 tractor units to run in the famous livery through his influence. The unit is pictured at the Musselburgh depot with one of the company's flatbeds, roped and sheeted, in tow. This trailer type remained in service with Pollock into the 1990s.
(Photo: John Henderson)

Back cover photograph
The Robson livery and name were so well known that United Glass decided to retain them when it bought the company in 1980. Robson's 2800s were perhaps some of the best known and most easily recognised of any operator in the country. All bearing a Border name, this one was maybe the company's 100th DAF unit. *(Photo: DAF Trucks Ltd)*

Cover design and book layout by Liz Whatling
Printed and bound in China

Contents

	Page
FOREWORD by George Bennett	4
INTRODUCTION – a brief history of DAF	6
DAF 2600 & 2200 – raising the bar	8
DAF 2800 – blazing the trail	10
DAF 3300 – what's in a name?	82
DAF 3600 – a true flying Dutchman	126

Foreword

By George Bennett
Editor, Truck *magazine 1987-89 & 1990-97*

Of the many trucks I drove as a professional driver in the 1970s and '80s, including Volvos, Scanias, a Ford Transconti and even a White Road Commander (of which the less said the better), the DAF 2800 remains my favourite. I am glad that Patrick Dyer's fine survey of the 2800 and its derivatives has given me a chance to revisit my own experience with it.

I first became aware of the DAF when I saw it in the pages of *Truck* magazine, in an advert for the 2800, featuring owner-driver Robin East, later of Rokold fame. It looked magnificent with its wide screen and huge, purposeful-looking cab. Then, in about 1977, my company, Cadwallader, bought out Macpherson's Transport, and with it a number of 2800s. I was driving a Volvo F88 at the time and so the big DAF cab seemed like a ballroom in comparison.

The first time I actually drove a 2800, however, was for Harrison International, a modestly sized Middle East operation based in Brierley Hill in the West Midlands. My first trip for the firm was a gruelling drive to Dubai in a less-than-wonderful Magirus Deutz that broke down several times, but for my second and subsequent trips I was promoted to a DAF 2800 DKS, one of five that Jack Harrison had bought new to service a major subcontract to Dubai and Jeddah. A year later, when I returned to better-paying European work for Cadwallader, I was given one of the ex-Macpherson DAFs, also a 307 bhp DKS.

Though the DAF didn't have the prestige of a Scania or Volvo in those days, it was a real driver's motor, which had almost everything right, from the steps upward. The view from the high-mounted cab and wide, flat windscreen was outstanding, while the commanding driving position was good for safety and the driver's self-esteem. The only let-down in early models were the narrow standard mirrors, and before I left on my first trip I had bought a second mirror for the passenger side, aimed wide to monitor the trailer in tight corners.

By the standards of the time, the power – and, in particular, the hefty torque – of the DKS were excellent, while the 13-speed Eaton-Fuller gearbox was a real professional's gearbox and a joy to drive. Despite the fact that the gearstick was a tall, hefty wand of an affair that stood up from the footwell, the shift was precise and light. For much of the time you could shift without using the clutch pedal, something not possible with the cumbersome ZF synchromesh box that replaced the Fuller in later models.

As a manufacturer, DAF has long had a reputation as the long-haul driver's friend, and the 2800 was the model that began it. The cab was about as wide as was possible within the rules and it was mounted high enough to allow a very low engine cover for its day. Moreover, the 2800 came with two bunks as standard, which was welcome if you were carrying friends or family. In most rival cabs, two bunks compromised the living space for a driver on his own, but DAF's designers had come up with a brilliant idea that allowed the top bunk to be set halfway up, giving an ideal height for a single occupant and enormous stowage space beneath it. Patrick rightly called this a 'reasonable perch for a third crew member' but it also made the DAF the cab of choice for drivers to gather in when stuck in some god-forsaken port or border, since there was space for two to sit in comfort on the bunk with another driver in each seat, while still leaving room to brew up with a stove on the engine cover.

I remember cooking a meal in the DAF cab with my friend Joe Berrington when we were parked up one evening on the outskirts of Amman. We were approached by a Jordanian soldier guarding some military installation, and after a few minutes' chat we invited him to join us for a brew in the cab, where we

had an unexpectedly interesting conversation about, of all things, Middle East politics. There wouldn't have been much room for such entertaining in any other contemporary cab, not even in Joe's Transconti.

The flexibility of the bunk design was further enhanced by the fact that both mattresses were removable, which was an advantage if you wanted to sleep outside the cab, or simply wanted to air them out. The bottom mattress was located flat along the cab floor, where it retained heat for a couple of hours after the engine had been switched off, making the bottom bunk the choice in winter.

Not everything was perfect in the early 2800s, however, and Patrick rightly points out that DAF provided thicker curtains and, above all, better insulation in later versions. My R- and T-reg 2800s had nothing overhead but a thin plastic lining under the metal roof, which made it very cold in winter and provided a constant temptation to run the engine at night. Nobody needed the cold rear window of the early designs, which was first double-glazed and finally eliminated in the steady improvements that DAF introduced over the years.

Another little-known drawback of the 2800 was the oil filler. It is true, as Patrick describes, that daily checks were easy to make under the two-piece opening grille, but he does not mention the unofficial accessory that most long-haul 2800 drivers carried behind it. This was necessitated by the oil filler pipe, which presented a vertical opening that was impossible to reach with a can. Instead, many of us carried a two-litre water bottle with a panel cut from one side that acted as funnel if you pushed the neck horizontally into the pipe.

I only once came across DAF's specialist long-haul version of the 2800, when I ran from Munich to Syria with an owner-driver and his Supercontinental DKS.

Despite the fact he had a kitchen under the bunk, we never used it, for two reasons. The first was that it was a pain for him to move all his sleeping gear out of the way, and the second was that the weather was so warm we both preferred to use the cooking equipment in my trailer box, and sit outside on picnic chairs.

By the time DAF rolled out the 3600 ATi and the legendary Space Cab, I was working as a truck tester and journalist for *Truck* magazine. I remember attending the launch to journalists and enjoying not only the vehicles themselves but the memory of driving early versions all over Europe and the Middle East. All the drawbacks of the early models had been addressed, the new cabs were well insulated, all the windows behind the driver had gone and the ATi engines put even the DKS in the shade. The only thing I missed was the constant-mesh 13-speed Eaton box, which had given away to synchromesh, but you can't stand in the way of progress.

GEORGE BENNETT
March 2010

Introduction

A Brief History of DAF

DAF's extraordinary rise to prominence among Europe's heavy truck manufacturers is perhaps all the more remarkable given the company's comparatively late start as a producer of powered commercial vehicles.

Hubertus (Hub) Van Doorne was born in 1900 and, while not an academic star, he showed from a very early age a sound understanding of engineering principles and an imaginative approach to mechanical solutions. Forced out to work at the age of eleven following the death of his father, Hub had a number of jobs and false starts in business, not helped by the intervention of the Great War between 1914 and 1918. However, Hub finally established 'Van Doorne's Machinefabriek' in 1928 with the help of a wealthy backer, the owner of the local Valk brewery, and Hub's younger brother, Wim.

The 40, 50 and 60 series trucks were typical of the products produced by DAF in its early years as a manufacturer of commercial vehicles. Simply engineered and well made by the Dutch workforce, they forged a place in the market for the newcomer. The cab was of a distinctive, clean design with good, aerodynamic properties. The factory could also supply a bare chassis with just the front panel for coachbuilders. Note the early adoption of angled panes at the edge of the windscreen, a feature that would endure until 1987, and the clever mounting of the wiper arms through the glass. (Photo: DAF Trucks Ltd)

Wim, born in 1907, was as gifted a businessman as Hub was an engineer and the pair made a dynamic partnership. The business quickly took off with the small engineering shop churning out all manner of fabricated products, many for the brewery itself. The initial loan was soon repaid and a new factory replaced the original small workshop. It was now 1930 and despite the Great Depression the brothers identified a new and significant diversification into the manufacture of trailers. To mark this new direction the company's name was changed in 1932 to 'Van Doorne's Aanhangwagenfabriek', literally 'Van Doorne's Trailer Works' and if taken as an acronym, DAF.

DAF trailers soon generated a good reputation, featuring as they did innovation after innovation, most of which resulted in major weight savings over rival designs. It was also at this time that DAF developed a revolutionary container system for use on the Dutch railway network. Complete with specially designed handling equipment, the system was a great design and was adopted by other countries including France, Germany and Belgium.

The Second World War and occupation saw the DAF factory taken over and production shifted to support the German war effort. During this time the brothers worked on their ambitious plans for the future once hostilities ceased.

In recognition of a new post-war start the company's name was changed again in 1948 to 'Van Doorne's Automobielfabrieken NV'; crucially the initials remained the same, and production of commercial vehicles started in 1949.

This put DAF some thirty years or so behind the established manufacturers, but it did mean the company was not tied to previous ideas and practices. It is interesting to note that from the beginning DAF favoured the forward-control layout, while others still persisted with the cumbersome and less ergonomic normal-control designs.

The first signs of success came in 1950 with the introduction of the popular 40, 50 and 60 series of rigid and tractor chassis, produced in the company's new, purpose-built factory in Eindhoven, featuring Hercules and Perkins power plants, the latter a diesel. The range utilised weight-saving techniques developed for the trailer business, including welded chassis frames. Large orders for trucks were placed at this time by the Dutch military, which was busy re-equipping, and this increased production significantly.

By May 1955 the factory, now extended to 60,000 square feet, produced its ten-thousandth truck chassis and a new range, the 1100, 1300 and 1500 went into production. These were important models for the company as they featured the Leyland O.350 engine marking the start of an important period of co-operation between the two companies. At first these engines were supplied direct from the UK,

With the Dutch military busy re-equipping itself in the early 1950s, DAF found itself producing trucks of 4 x 4 and 6 x 4 configuration. The company's growth as a commercial vehicle manufacturer owes much to the stability that these orders brought during this period. (Photo: DAF Trucks Ltd)

but from 1957 DAF built its own version under licence in its new engine plant at Eindhoven.

The requirement to handle higher weights was met in 1957 with the introduction of the big 2000 series. This was a truck of serious capabilities equipped with a licence-built version of Leyland's 11.1 litre O.680 engine. This was enormously significant as it was this engine that would ultimately lead to DAF's superb 11.6 litre 'K' series design some ten years later.

In a further move towards manufacturing independence, DAF opened its own axle production facility in 1958. This bold step brought the company closer to the point of total control, as enjoyed by rivals such as Volvo and Sania, and left the gearbox as the only major component to be bought in.

The end of the decade brought DAF's first turbocharged engines and saw more new models with the 1600 replacing the 1500 and a new series, the 1800, being introduced.

DAF had enjoyed a remarkable ten years, but the next decade was going to be extremely important for the Dutch manufacturer.

DAF 2600 and 2200
Raising the Bar

By 1962 DAF had been producing trucks for just thirteen years, but they had used those years wisely, steadily refining and improving their products and, with every new type introduced, had climbed another rung up the weight category ladder.

At the start of the decade the big hitter in the line-up was the 2000 DO series of forward-control trucks and tractors. The range featured a strong-welded chassis, DAF's evolved version of Leyland's 0.680 engine, constant or synchromesh gearboxes and a capable 10 tonne axle. The range also saw the first sleeper cab to be offered by DAF, a clear sign of the way international routes were developing. Importantly, the mechanicals proved very reliable in service, especially the 0.680 that endeared them to operators and drivers alike.

It was this, the heaviest of DAF's ranges, which was to form the mechanical basis of the company's most important model to date, the 2600. However, it was not the mechanical aspect of this truck that was to cause the biggest stir; instead it was the remarkable cab. Gone were the old sloping profiles and rounded shapes of previous types. In fact, the only visual DAF trademark left to link the new vehicle to the past was the retention of angled panes at the side of the windscreen. The 2600 was a radical departure for DAF, a handsome four-square design that was ultra-modern in appearance and possessed a clean functionality befitting the start of a thrusting new decade. Indeed, this was the space age.

The interior was no less remarkable. Light, airy and spacious it was finished to an extraordinary high level for a truck and offered drivers the kind of luxury usually only associated with expensive cars at that time. With its powerful engine and strong performance, optional sleeper cab, comfortable seating, good insulation and heating, the 2600 was at the forefront of a new breed capable of undertaking true long-distance work, the TIR truck.

The 2600 was updated throughout its fourteen-year career, most notably in 1964 and 1969, when significant improvements were made to the – already good – cab. In 1968, the old Leyland 0.680 officially became the DAF 1160 (K series) of 11.6 litre capacity, the engine which would ultimately go on to power the 2800. At this point the 2600 became equipped with the DKA version featuring tuned-port ram induction offering 230 hp. A 13 tonne axle followed and, in 1972, the DKB 1160, which saw power rise to a prodigious 304 hp making the 2600 capable of 42 tonne operation.

If the 2600 had a failing, it was that the cab was of the fixed variety. All servicing had to be carried out through access panels, and an engine change would require the cab to be removed completely. While insulation was very good, it would have been much better for the driver without all those panels being present. To be fair, the 2600 had bridged the gap between fixed and tilting cabs, and had easily made the best job to date of the former types when it was introduced in 1962. DAF's engineers were acutely aware of the problem, and by the end of 1969 they were ready to announce their first tilting design, the F218.

The F218 cab was to be the crowning feature of a new range comprising the F1600, 1800, 2000 and 2200. This was an ultra-modern cab, with more than a passing resemblance to that of the 2600. The design was modular but at the time truly innovative, so that DAF could create numerous different types of varying widths, heights and depths from a common set of panels and frames. Tilting was a manual operation, much like Volvo's F88, and was only possible after both an interior lever and exterior safety catch were released. The action was assisted by two torsion bars, allowing one-handed operation even for the 640 KG sleeper version. The cab would tilt to a full 60 degrees before mechanically locking, but all daily checks could be carried out with the cab down. Rubber pads and shock absorbers provided suspension. The driver's environment was superb, with 100-degree opening doors giving easy

The 2600 caused a major stir when it was introduced in 1962. The tall, handsome cab concealed mechanicals of proven reliability from the old 2000 series and included DAF's constantly developed versions of Leyland's O.680 engine. Interiors were sumptuous and set new standards for the day, offering the driver a true home from home while travelling between countries. The model enjoyed constant development with two major revisions in 1964 and 1969. Power peaked at 304 hp with the introduction of the DKB 1160, the installation of which required the cab to be raised by approximately 50 mm. (Photo: DAF Trucks Ltd)

The F218 tilting cab first appeared in series production in 1970 and formed the basis of the larger version, which would debut on the 2800 range some three years later. As the design was modular, combinations of standard panels could be used to create many different variations. DAF also offered a fixed version from the same panels, known as the 'Windscreen' model as it ended behind the front pillars, for use as the base for Pantechnicons and municipal vehicles. Wyko operated this FM2000 16 tonner under TIR permits between Birmingham, Stuttgart, Amiens and Liège. It was fitted with the DAF 8.25 litre engine producing 163 bhp and a six-speed, ZF gearbox. The truck featured tilt bodywork by Silverdale and an Acoma tail lift. (Photo: DAF Trucks Ltd)

access to a fully adjustable sprung seat, affording a superb view of the road through the amply glazed cab.

It was to be a marriage of the 2600's proven mechanical components and a bigger version of the F218 cab which was to form DAF's next remarkable truck, the legendary 2800.

While the 1960s had been an epic decade for DAF with the introduction of several milestone trucks, the company was cash poor. By the start of the 1970s the huge development and tooling budgets required for the engine and cab programmes had taken a big toll on the finances. Luckily, two important deals were signed in 1972. First, DAF joined the 'Club of Four', a co-operation agreement with Volvo, Saviem and Magirus to develop light trucks up to 12 tonnes. Second, and more importantly, DAF was offered a transatlantic lifeline by the US giant, International Harvester, IH, then the number-one producer of heavy trucks in the USA, was desperate to expand into Europe and other new markets around the world, but unfortunately its home-market products proved unsuitable for the task. The deal, which gave IH a one-third share in DAF, brought in the much-needed cash in exchange for co-operation on design and technologies. Although design and development for the 2800 programme was well under way when IH came on board, it did outwardly represent the first fruits of the new relationship.

A further boost came in 1976 when the DAF car division was sold to Volvo allowing the company to focus entirely on commercial vehicles.

DAF 2800
Blazing the Trail

The outstanding 2800 series was introduced late in 1973 at the Frankfurt Show to a rapturous reception. DAF was already riding high on the success of the 2600 and the superb F218 tilting cab of its lower-weight machines, but the 2800 catapulted it into another league altogether. At a stroke it was now a producer of a super premium truck capable of taking on the best from rivals such as Volvo, Scania and Mercedes.

The new truck featured a high-tensile steel chassis, which, in a move away from DAF's previous practice, was riveted and bolted rather than welded. The side members were flat-topped to assist the mounting of bodywork on rigid examples, while tractors came with a generous flange for fifth-wheel location. The rigid examples were provided with a sturdy rear cross member pre-drilled for a drawbar coupling. Somewhat uniquely, DAF, still a significant trailer manufacturer, could also supply the truck's bodywork and a matching trailer for drawbar operators.

Front and rear axles were both of DAF design and manufacture. Drive axles were available with either a 10 or 13 tonne capacity for 4 x 2 configurations, while 6 x 4 bogies were offered with 16 or 20 tonne capacity. Parabolic springs were standard for tractors and could be supplemented with anti-roll bars if necessary. Rigids were equipped with a semi-elliptic spring and double-acting shock absorber arrangement to which could also be added anti-roll bars.

At the heart of the 2800 was the excellent DAF 1160 six-cylinder engine. As originally offered, this was available as the DKA, DKTD, DKT and DKS providing four different outputs to suit operators' requirements. For the UK this choice was rationalised to the 248 bhp DKTD and the 307 bhp DKS. The lower powered DKTD ran to a maximum of 2200 rpm and used relatively mild turbocharging to give a useful 650 lb/ft of torque, placing it nicely in F88 and LB110 territory.

These characteristics made it ideal for UK haulage and because of the engine's low-stressed nature it was consequently to prove highly reliable and economical with great longevity. The DKS was highly significant as it represented the first use of turbocharging and charge cooling (intercooling), by a European truck manufacturer. DAF engineers knew that reducing the temperature of the compressed air from the turbocharger would bring positive benefits and had been running trials of the system using a modified 2600 tractor since 1969 with great results.

The final system adopted for the DKS used an air-to-air unit mounted ahead of the main radiator to effect cooling of the compressed air by around 50 degrees centigrade. When cooled, more air could be forced into the engine's combustion chambers and with better metering a near perfect air-to-fuel mix was possible. The result was a boost in power to 307 bhp and a prodigious rise in torque to 855 lb/ft, which was available at a lowly 1450 rpm.

Gearboxes were by ZF with a six-speed unit for the DKTD, available with a splitter to double the ratios, and a thirteen-speed type for the DKS. While both constant-mesh and synchro boxes were offered, the former were often preferred by drivers because of the fast, positive shift and by owners for the extended service life.

The 2800 package was completed by the introduction of the F241 cab. This was based on the F218 modular panels, but was some 200 mm wider thanks to a central filler panel unique to the F241. It was also mounted higher with door bottoms level with the top of the wings, themselves a flush fitting, rubberised item unique to the F241.

The standard cab was the super-comfort twin sleeper – the day cab was an option – and it came well equipped with excellent Bostrom seating, a full suspension type for the driver and a reclining version with arm and head rests for the passenger. The bunks were generous for the time, at 600 x 1900 mm, and featured a contoured edge to prevent the occupant rolling out. The upper bunk could be stored in a lowered position above the bottom one, providing a reasonable perch for a third crew member. Instrumentation was clear and logically laid out in front of the driver in an angled panel that protected it from reflections. The roof was heavily domed giving good headroom, and space was generous with good provision for storage including a useful centre console. The cab was finished with a new, deep full-width grille of black plastic across which were prominently placed the letters D A F. All daily checks could be done via this and the smaller grille below. Cab tilting, when required, was assisted by a hydraulic pump and a full 70-degree tilt angle was available – 68 degrees according to some testers. Five standard colours were available: red, blue, green, yellow and grey, with a further four on the option list.

In 1975 DAF followed an industry trend and announced the 'Supercontinental' option for the 2800. This saw the F241 cab kitted out with all the accoutrements thought desirable for those drivers engaged on ultra-long jobs like the Middle East run. A true home from home was provided with a sink, cooker, fridge, table, water storage and shaving mirror all contained in a purpose-built unit beneath a hinged bunk. A wardrobe and toolbox were also provided. Tinted and laminated glass, external sun visor, air conditioning and an independent cab heater all added comfort, while mesh grilles protected the lights and radiator. The base vehicle for this package was usually a left-hand drive DKS tractor with a 3.1 m wheelbase and cast-iron fifth-wheel coupling.

1975 also saw the introduction of an air-suspension option for models fitted with the 10 tonne rear axle. This was a twin-bag system by Brunninghaus AG, energised from its own air supply with bags mounted on a sturdy frame behind the axle. The pivot point was ahead of the axle near the normal front spring hanger location and was linked to the air bag's frame by a double spring of which the lower and more substantial one dog-legged under the frame, while the other served as the upper rear locator.

A special heavy-duty 2800 was announced in 1977. Aimed at the export market beyond Europe, in particular the Middle East, this range featured 6 x 4 layouts for trucks and tractors installed in beefed-up frames. Key hardware, such as air-tanks, was relocated into or above the chassis rails. A double-skinned roof, air-conditioning, robust bull-bar, light guards and Trilex wheels were often fitted as well.

The first major change to the 2800 range came in September 1979 with the announcement of the super economy DKSE version of the 1160 engine. Despite being noted for their frugality, the fuel efficiency of the DKTD and DKS were surpassed by this new, low-revving development. Like the DKS, the new engine was both turbocharged and charge-cooled, but with a lower maximum running speed of just 1800 rpm, the DKSE produced only 276 bhp.

However, more air could be supplied at slower speeds thanks to a special turbocharger and a different governor setting which increased torque remarkably. The DKSE would deliver 929 lb/ft at 1300 rpm, while the DKS could only manage 855 lb/ft at 1450 rpm. In theory this meant less gear changing was required, but the optimum rev range – green band – was narrow, only around 350 rpm, so a 16-speed ZF box was specified giving a gear for every situation. This was a synchromesh unit which, combined with the 4.49:1 rear axle, allowed a maximum of 59 mph. While this was fine for continental-spec vehicles running at a legal 50 mph, the set-up would be no good for UK operators who could at the time run at 60 mph.

Consequently, UK DKSE-equipped vehicles were supplied with a Fuller RTO 9509A 9-speed unit allowing a maximum of 66 mph at 1800 rpm and a maximum gradeability of 1 in 4.2 for a moving vehicle, or 1 in 4.8 for a standing start. With a torque curve that dropped away suddenly below 1100 rpm and the narrow economical rev range, the new truck required a very precise driving technique to get the best from it, so DAF offered tutorial sessions to salesmen, operators and drivers.

Shortly after, DAF also introduced the VISAR driving aid, which used a black box monitoring device with a three-light system to advise on the best gear selection. While slightly annoying to a competent driver, it did take the guesswork out and a VISAR-equipped DKSE could be a very economical machine indeed.

Both these developments coincided with the introduction of a face-lifted cab, known as the second generation. Outwardly identical, save for new badging, the cab was transformed inside with warm materials for bunks and seats, heavier curtains, a redesigned dash with rationalised switches and a smaller, padded steering wheel. Insulation was increased in the floor, walls and roof, and – answering a serious criticism of the original design – the rear windows were now double-glazed.

Combined with a new colour scheme of brown and beige, these changes created a superb driver's environment. Revisions to the electrical system brought the relay/fuse box to a new location between the seats on the engine cover, making fault-finding and servicing an easier affair. There was a new multi-function stalk for horn, wipers, hi/lo beam and indicator operation. Outside, a new one-piece plastic battery cover with quick-release fasteners was installed, eliminating the risk of shorting from the previous metal cover.

The last significant change to the range was the introduction of engine changes under the ATi (Advanced Turbo Intercooling) banner, but as ATi also affected other, yet to be introduced models, we will look at these in a later chapter.

Although the 2800 range was introduced in autumn 1973, it was not until the following spring that UK deliveries started. This early example, an FA2800 (A-denoting a rigid chassis), was still gainfully employed by Visbeen's UK operation, when over ten years old. Its drawbar configuration and left-hand drive layout suggests that it may have been transferred from the ranks of Visbeen's Dutch fleet, where the type was common. The UK operation, however, relied mostly on tractor/trailer combinations, so possibly it was kept on beyond the usual six-year changeover period, as its configuration was so versatile.
(Photos: Dave Wakefield & DAF Trucks Ltd)

More 2800s started to appear in the UK with the new N registrations of 1974-5. This excellent example was one of UKON Trucking's earliest and was virtually brand new when photographed at the Imperial War Museum, Duxford with the wings of the B17 Flying Fortress 'Mary Alice'. The WW2 bomber was brought back from an airfield just outside Paris in an operation involving three of UKON's fleet, this new FT2800 and two Volvo F88 240s. The fuselage travelled in two sections loaded on flats behind the Volvos, while the wings were loaded onto an extending Broshuis three-axle trailer leased from TiP. Two years later UKON transported the museum's Short Sunderland flying boat from France in a similar operation. (Photo: Pieter Kroon)

HJB 600N, 'Miss Alicia', was an early star of DAF's advertising campaign for the new model featuring in dealer adverts alongside trucks from other well-known operators such as Jekinsons, Yardley and Cabmont. The smart yellow and black unit is pictured in the summer of '76 while negotiating Poplar in London. (Photo: Joe Donaldson)

Econofreight was an early customer for DAF's range-topping FTT2805 tractor unit. The FTT2805 (T for tractor, T for 6 x 4, 05 for heavy duty) featured the DKS engine giving 307 bhp and a thirteen-speed Fuller RTO gearbox driving DAF's own 6 x 4 back end. The load in this case was one of two 'Air Heaters' which the company moved from the manufacturer, McTay Engineering, in Peterlee, Co. Durham to Monsanto Chemical's plant in Seal Sands some fifteen miles away. At 51 tonnes apiece and measuring nearly 53 ft by 18 ft, the move was a challenge taking three hours. The five-axle King trailer was extendable and was usefully equipped with steering on the last two axles.
(Photo: DAF Trucks Ltd)

Ignore the misleading registration, this early unit was bought second hand by Cammack around 1979 and had previously done a two-year stint on Middle East work which had taken quite a toll. Cammack undertook an extensive renovation before applying its smart livery, with customary 'Colne Valley' name, and put the unit to work pulling tilts all over the UK. Note the narrow mirrors fitted to the early 2800s.
(Photo: Jim Cammack)

This smart Greek unit was at least ten years old when snapped dropping into Dover at the start of its long trip home. Despite being a fairly small market for heavy trucks at the time, DAF's 2800 proved a popular choice for Greek operators and drivers. Note the slightly different front wing treatment of this example – which was possibly a local replacement part as it was also seen on other Greek 2800s – and the addition of Volvo F10/12 indicator assemblies.
(Photo: Dave Wakefield)

It was the economy and reliability of the DAF product, combined with the superb back-up systems of DAFaid and ITS (International Truck Service) that made Herreveld International such a loyal DAF customer. In 1976 the company placed a massive order, approaching £200,000, for thirteen new DAF units. The order mostly comprised the smaller FT2300s but included one heavyweight FTT2805 too. Much of the company's work was with TIR tilts, such as this one hauled by an FT2800, to destinations including Spain, Portugal, Switzerland and Yugoslavia. *(Photo: DAF Trucks Ltd)*

This 1977 unit of Welsh operator Evans & Williams looks superb with its beautifully applied livery. The F241 cab of the 2800 featured ideal break points in its panelling and made a perfect canvas for such a scheme. This FT2800 was fitted with the DKTD engine giving 248 bhp, ample power for this 35 foot single-axle box van and its 'high cube' load of shoes. Clarks bought DAF units themselves in 1985 when two 28 tonne FT2100 units joined the fleet.
(Photo: Mark Lester)

JY Paxton was an owner-driver who operated this fine left-hand drive FTT2800 DKS on the Middle East run. At the time of the photo, a young George Bennett, who was on loan from his employer while he waited for a load of his own to Lahore, was driving the unit. The truck looks very purposeful coupled to this tilt as it waits at the roadside in Essex en route to Dover. Note the air conditioning unit mounted on the roof.
(Photo: George Bennett)

Dowty ran five FT2800 tractors like this with the DKTD and ZF six-speed combination. The 248 bhp DKTD was about right for UK 32 tonne operations and, being well within the ultimate capabilities of the 1160 engine, proved astonishingly reliable in this low-stressed state. Dowty's trucks were employed on delivering the company's roof mining supports throughout the UK and Europe using semi-low loader trailers. The units were supplied under a contract hire scheme from BRS. Dowty later merged with Gullick Dobson and became Longwall International which, in turn, became part of Joy Mining Machinery.
(Photo: DAF Trucks Ltd)

DAF engineers followed the trend set by manufacturers like Volvo and designed a snorkel-style air stack for the 2800 to gulp cool fresh air from over the top of the cab. However, unlike Volvo's F88, the 2800's air stack would remain on the right-hand rear of the cab irrespective of whether the truck was right- or left-hand drive. European left-hookers were fitted with a straight item, but to preserve the over-the-shoulder view of the driver, UK trucks got this neatly cut-and-shut item that side-stepped the window, as illustrated by this smart example in UKON's excellent livery. *(Photo: Pieter Kroon)*

One of UKON's specialities was the door-to-door delivery of forklift trucks. New machines ranging from 2 to 20 tonnes were collected from manufacturers such as Lancer Boss in Leighton Buzzard, JCB in Uttoxeter and Hyster in Irvine and delivered directly to the customers throughout Europe. Some were transported on dedicated trailers like this four-in-line low loader, while others jostled for space as part of groupage loads carried in tilts. *(Photos: Pieter Kroon)*

Owner-driver Dave Dickson kept his smart FT2800 busy on a diet of continental trips and UK drops with its dedicated tilt. Dave's early livery, like many DAF operators, used fine pin striping of the cab panels to great effect. The well-travelled unit featured several factory options such as the sun visor, illuminated headboard and front bumper with spot and fog lights installed, well over £200-worth in those days. *(Photo: Adrian Cypher)*

An FA2800 makes perfect sense as the base vehicle for this high-volume load of racing pigeons (?) spotted on the move down the M4. The 2800 range was designed with just as much emphasis on drawbar operation – such was the popularity of that configurations on the continent – as it was with tractor units. As such, the 2800 chassis in rigid form was completely flat and featured a sturdy rear cross member for the towing hitch. Note that this early UK-registered example with left-hand drive is fitted with much later corner deflectors. *(Photo: Adrian Cypher)*

This FT2800 unit, operated out of Dunblane by Craig's Bulk Freight, shows again how well the big cab suited the application of a traditional UK livery. Running empty, or part loaded, this example was obviously capable, given its two-plus-three configuration, of operating at 38 tonnes. As the DKTD would have been at full stretch at this weight, it is reasonable to assume that this unit was probably fitted with the 307 bhp DKS engine, which offered over 200 lb/ft more torque. This unit has been fitted with the later under-bumper airdam which became available in 1982.

(Photo: Marcus Lester)

It was the closure of Christian Salvesen's north-east facility that led four redundant drivers to strike out on their own and form Bon-Accord Transport. MAN 16.232s were originally considered but finally rejected because of the column change, which left the door open for DAF. Five FT2800 tractors were bought, the fifth to be piloted by a driver supplied by Vic Mathers. The units were fitted with the DKTD engine, ZF six-speed gearbox with splitter and DAF's own hub reduction axle, the 2699, with the standard ratio of 5.53:1. This unit is pictured at the company's usual tipping point, London's Smithfield meat market. DAF fully understood the importance of the Scottish market which accounted for 10 per cent of UK heavy truck registrations annually, so it built a dedicated facility at Westmains, Grangemouth with sales, service, parts and account facilities.
(Photo: DAF Trucks Ltd)

With the F241 cab being offered in the twin-sleeper format as standard, the day cab was actually a cost-reducing option for the 2800 buyer saving nearly £300 on the purchase price. This example was operated by Brain Haulage of Essex and could have been one of two such units operated on a day-hire basis from the company's depot in Grays. There were many other DAF units, both DKTD and DKS, in the 350-strong Brain fleet, many operating in the livery of customers such as ACT and CGM (UK) Ltd.
(Photo: Marcus Lester)

With its roots in the late 1950s, John Beynon Evans' business not only covered general haulage and container movements, but also offered extensive maintenance and servicing facilities to such organisations as Christian Salvesen and BOC Transshield. The DAF FT2800 pictured was the company's first and was bought only after close comparison with the competition's vehicles. The company's findings were that the DAF offered a better all-rounder at this weight category and placed special emphasis on the superb back-up and parts support. The DKS-powered unit performed admirably, servicing a contract to Bell Lines. *(Photo: DAF Trucks Ltd)*

Oakleigh Animal Products of Ascot started out as a door-to-door delivery service for domestic dog food in 1950. By 1978 the company was operating at the other end of that particular food chain and was collecting the raw materials from abattoirs around the country and delivering it to the pet food manufacturers for processing. Five DKTD-powered FT2800s were employed on this work, including three to special order equipped with day cabs. OAP also had customers abroad, but the units equipped with sleeper cabs probably serviced these. (Photo: DAF Trucks Ltd)

This FT2800 DKTD was bought new by Cammack and was finally picked over the more expensive F88 which driver Jim Cammack wanted. However, once he got behind the wheel of the big DAF he was soon swayed and never looked back. Jim drove the unit, named 'Colne Valley Empress', for the next three years, mostly pulling tilts throughout the UK for Trailermann. Jim's only criticism was that the gearbox didn't work well when cold. Note the stylised lettering used for the DAF name on early units. (Photo: Jim Cammack)

While UKON definitely had its specialist fields of activity, it was always willing to take on different work and, enjoying a close working relationship with TiP, would rent in the appropriate trailer to perform any given task, as is seen here with one of the company's DKS units rather unusually coupled to a fridge van. (Photo: Pieter Kroon)

UKON's fleet consisted of Volvo, Scania and DAF tractors with the odd Ford D-Series thrown in for local work, but with the company's strong Dutch connections it is little wonder that the majority of heavy units were FT2800s. The DAFs were found to be powerful – most were DKS, reliable and, above all, backed up by a superb support network. Here, a selection of new S-registered examples is assembled for the camera at the Hoddesdon yard. (Photo: Pieter Kroon)

It was slightly unusual to see a Russell Davies truck with a roped and sheeted load rather than an ISO container. Russell Davies was particularly keen on the reliability record of their DAF units and hadn't suffered one breakdown by 1979. However, driver acceptance was also of importance and the company's FT2800s with DKS power scored high here too. In a fleet with many drivers' trucks, such as the superb LB111, that could only be good news for DAF. Note the locker mounted under the batteries on this example and the mesh protecting the spotlights in the bumper, which could be part of the 'supercontinental' package. (Photo: DAF Trucks Ltd)

Brian Swindell's background as an agricultural engineer set him up well for a career in road haulage and gave him a good appreciation for no-nonsense engineering. As the DAF 2800 range was built to such a high standard and on solid engineering principles, it was obviously attractive to Brian. Many FT2800s with DKS engines were employed on various contracts including a large haulage and storage operation for CP Ships.
(Photo: DAF Trucks Ltd)

In 1979 Jack Harrison's fleet numbered eleven units: five Magirus-Deutz 310s and six DAF FT2800s, five of which were brand-new T-registered examples. The DAFs were part of a fleet expansion made possible by a large subcontract job for Spiers & Hartwell which kept the Dutch units busy running between Glasgow and the Middle East delivering equipment for desalination plants. The three are pictured in Jordan in the spring of 1979 and all are standard DKS units.
(Photo: George Bennett)

By 1979, Cleveland Tankers had over sixty DAF trucks in its fleet, including the first of the new 2100 models to be put on the road. Formerly known as Rankin Tankers, the company was engaged on the transport of chemical and petroleum products with its specialised tanker fleet. Most of the company's big DAF units were equipped, like this one, with the DKTD engine as performance was perfectly acceptable at 32 tonnes. (Photo: DAF Trucks Ltd)

Topping the tanks Middle East style may not have been a glamorous affair, but the diesel was certainly cheap. Always noted for its economy, the big DAF's lowly fuel consumption, even in testing conditions such as this, made a valuable contribution to profit margins for operators. Despite being a typically tough haulage man, Jack Harrison was sentimental about his trucks, especially the DAFs, and would often drive ahead of a departing convoy to admire them from the vantage point of a motorway bridge. (Photo: George Bennett)

This FA2800 was the first 2800 to be put to work as a car transporter in the UK. The rigid vehicle ran with a then unique drawbar trailer manufactured in Preston by Municipal Trailers. This close-coupled item had just one pivot point and so behaved like a conventional artic when reversing. As the cars could be driven through the trailer onto the truck, even onto the top deck, uncoupling was generally not necessary. Contracted to deliver Datsun cars from the ports of Middlesbrough and Southampton to destinations throughout the UK, Car Removal Ltd opted for the more powerful DKS and were impressed with the vehicle's high-speed cruising ability. (Photo: DAF Trucks Ltd)

Despite looking for all the world like a racing car transporter of the period, this handsome FT2800 DKTD and step-frame box van combination actually served the parts distribution for that most romantically named automobile manufacturer, Alfa Romeo. EMS of Dunfermline ran the contract with this unit and a smaller DAF 2000. The larger truck performed the trunking role, covering up to 2,000 high-speed miles every week. Note the large catwalk diesel tank fitted on this unit. (Photo: DAF Trucks Ltd)

Thomas Allen Ltd was founded way back in 1854, but much like Robert Armstrong's tanker operation, was absorbed into the Road Tank Group of P&O. Thomas Allen had a rich heritage in haulage and over the years had developed a specialised service for moving liquid loads. He could handle anything from chemicals and fuel to latex. By the late 1970s, DAF trucks figured largely in the mixed fleet, which included rigids and artics. As well as the big 2800s, Thomas Allen also operated examples from the 2200 and 2300 ranges. This FT2800 was a 248 bhp DKTD unit. *(Photo: DAF Trucks Ltd)*

This FT2800 DKS with day cab was the first DAF unit to be fitted with an automatic transmission. Developed by ZF and Fichtel and Sachs, the system was designated the Transmatic. Using a torque converter and clutch in conjunction with a six-speed synchromesh box, the Transmatic was actually semi-automatic. With a torque conversion factor of two and a half times, the system gave great flexibility and this reduced the number of gear changes required, significantly extending clutch life. *(Photo: DAF Trucks Ltd)*

Following good experiences with the F88, UKON bought early examples of Volvo's new F10, operating a batch of S-registered examples. However, it continued its mixed-fleet policy and added more DAF FT2800s. This 1979 unit is coupled to a new 'Jumbo' tri-axle step-frame tilt. With the growing success of UKON's groupage service to Europe, trailers like this proved extremely useful to maximise volume.
(Photo: Pieter Kroon)

Bon-Accord's founding members, Bert Gilbert, Dennis Pickard, Ian Whyte and Alistair Wright, stayed loyal to DAF. BRS 365T was another FT2800 DKTD unit which followed the original batch into the smart blue, white and tartan livery. Because it was established specifically to haul the finest Scottish beef south to market, Bon-Accord placed much importance on reliability and back-up for its trucks. It also specified the best trailers it could, such as this chassis-less example by Gray and Adams of Fraserburgh. Constructed from polyester and aluminium, these trailers were light, strong and thermally very efficient.
(Photo: DAF Trucks Ltd)

This truly fine example of an FT2800 tells us, courtesy of the new and prominently displayed badges introduced in 1979, that under the cab is the 1160, six-cylinder engine in DKS form giving 307 bhp and 855 lb/ft of torque. Note the additional 88-gallon tank fitted to the offside of this unit and the relocation of the airtank to accommodate it. *(Photo: Marcus Lester)*

It was in 1979 that DAF's marketing department realised the error of its ways. For six years DAF had been producing 2800s with no identification at all, so DKTD-powered trucks could have been DKS and vice versa. It was decided that both the range and model identification would be prominently displayed. DKTD-equipped vehicles would get the word 'Turbo' below a 2800 badge while DKS examples would also get the word 'Intercooling'. Despite having lost its badges, this well-worked FT2800 does illustrate clearly the new area allocated for them in the corner of the lower grille. Note that on the earliest of these newly badged trucks, the old stylised lettering was still used for DAF on the main grille. *(Photo: Marcus Lester)*

Although this interesting Greek unit displays the 2800 legend, it was probably retro-fitted to this pre-1979 example as the badge has been stuck on the original-style lower grille without the blanked area for model identification. The truck also features the much later corner deflectors and a very interesting roof spoiler. (Photo: Dave Wakefield)

Of course badges, when tampered with, can become confusing. For example, this left-hand drive unit cannot possibly be just intercooled and not turbocharged also, as it is the compressed air from the turbo which ends up in the intercooler. However, it does look grand in its slightly tired state, a true TIR contender with many international miles under its belt, no doubt. (Photo: Dave Wakefield)

'This is it, the real thing,' the first FT2800 to appear in Coca-Cola colours in the UK. The DKS unit with day cab was based at the Southampton depot, which covered a broad area of 1,200 square miles along the south coast. At the time, Coca-Cola Southern Bottlers Ltd had just amalgamated with Cantrell and Cochrane. Note that the DAF lettering on the main grille is no longer the old stylised version. (Photo: DAF Trucks Ltd)

The DAF FT2800, an industrial triumph in an industrial setting. At the helm of Kentvale Transport was Kathryn Harrison. Unfazed by the male-dominated industry, Kathryn built up a smart fleet employed on UK and continental general haulage. Careful planning of back loads saw her DAFs operating to the optimum level possible. This DKS example shared the same spec as others in the fleet. DAF service and back-up were the key to the decision to buy, particularly the ITS scheme which allowed Kathryn peace of mind for trucks and drivers on the continent. *(Photo: DAF Trucks Ltd)*

EDP 774V was an enormously significant truck for concert tour experts, Edwin Shirley Trucking. The DKS unit was the company's first DAF and marked the change in buying policy that saw Volvo's products out and DAF's very much in, and the policy still stands to this day. It was an open invitation to manufacturers to supply a truck for EST's legendary calendar in 1979 that led to DAF salesman Ron Sinclair turning up with a yellow FT2800 DKS tractor unit, XDP 490T. EST were very impressed with Ron, so the truck joined the EST fleet for an extended demonstration term at the end of which the order for EDP 774V was placed. *(Photo: Dave Wakefield)*

As a young lad, the author was in awe of Rokold's smart fleet of 2800s. Painted white with a simple striped livery, the truck and trailer combinations were always immaculately turned out. The units were the more powerful DKS variant and usually had the sleeper's windows deleted or painted over, always the best look for the 2800. Robin East, Managing Director and founder, rated the big DAF's performance, reliability and back-up. (Photo: DAF Trucks Ltd)

The bulk tippers of All Seasons were generally loaded with coal, stone or grain and in the late 1970s were usually pulled by one of the company's FT2800 DKTD units. After trying many other makes, it was the DKTD-powered DAF that was found to be perfect for its operation. Note the relocated air tank which allowed the hydraulic reservoir for the trailer's tipping gear to be mounted on the off-side chassis rail. *(Photo: DAF Trucks Ltd)*

The upper part of a boat, possibly for the RNLI as the truck bears their flag, makes a most interesting load for this smart FT2800 DKTD. The operator obviously took great pride in the truck and adorned it with extras such as the vertical exhaust, chassis locker, fabricated rear wings, roof rack, catwalk tank and unique front bumper. *(Photo: Marcus Lester)*

DAF introduced improvements to the 2800's electrical system early in 1980. The most visual of these was the new battery cover which kept the batteries clean beneath a strong, quick-release plastic moulded item and eliminated the possibility of shorting the system through the build-up of terminal crust as could happen with the old metal cover. Other changes included a re-positioned junction box, new chassis wiring, new fuse and relay box, re-designed fascia panel, and a multi-function control stalk. *(Photo: Dave Wakefield)*

By 1980, DAF was quoting a slight increase in output for the DKTD of 252 bhp at 2200 rpm and 660 lb/ft of torque at 1500 rpm. Charter Roadways' latest DKTD FT2800 was so equipped and joined five other earlier units. The company converted to 2800s following a very successful period with a DKTD demonstrator in 1977. Charter Roadways provided vehicles and drivers for a number of well-known names including Ferrywagon Ltd, Eurofreight Ltd and B+I line. *(Photo: DAF Trucks Ltd)*

Dave Dickson's success as an owner-driver – mainly down to him being a thoroughly decent chap who was highly thought of by his fellow hauliers – allowed him to expand his business into a small fleet with more DAFs. This stunning unit, a DKS featuring his later livery, spent its time on a mix of continental work and block deliveries throughout the UK. *(Photo: Adrian Cypher)*

UCW 600V was the 2,000th F2800 to be sold in the UK. It was also the first V-registration DAF to take to the road. By now DAF held the number-two position for imports of heavy trucks over 14.5 tonnes and had 10 per cent of the 28 tonnes and over tractor market. This unit was the first DAF to join the 200-plus fleet of Blue Dart and gain the famous yellow and blue livery. Note that both a catwalk and catwalk diesel tank are fitted to this example. *(Photo: DAF Trucks Ltd)*

DAF introduced a new double-drive axle assembly, designated 2698T, in 1980. Manufactured at DAF's axle plant, the new unit replaced the old 2255T and offered several benefits, including improved cornering stability, better ground clearance and greater travel. The new unit was also designed to be largely maintenance free. A 2800 tractor fitted with the new axle was designated an FTT2816 – 16 represented the upgrade – previously a 6 x 4 tractor would have been an FT2805. However, in neither case would the truck be badged as anything other than a 2800.

(Photo: Marcus Lester)

This smart FT2800 DKTD was new to Tom Hughes and was the first truck supplied by DAF's new dealer facility in Wales. Well positioned next to the docks, Cardiff DAF Trucks dealt with all the sales and servicing requirements for South Wales, had extensive 5,000 sq ft workshops and held a large stock of spares. It was through keeping high levels of stock in strategic depots that DAF was able to maintain 94 per cent parts availability for the UK. *(Photo: DAF Trucks Ltd)*

The saying 'a picture paints a thousand words' is never truer than when presented with a superb photograph of a wonderful subject such as this, which leaves the author with the hairs on the back of his neck standing up and little more to say. Note the long-distance credentials of this FA2800 DKS festooned, as it is, with extra tanks and lockers. Fantastic! (Photo: Marcus Lester)

This fine machine was ordered and delivered to B Wallman before the new 2698T axle became available, making this unit officially an FTT2805. However, it still provided the owner with a fine prime mover for impressive loads such as this. Wallman had been around since the end of the 1930s and offered haulage and storage from its Manchester base. The fleet of forty or so trucks included several different models from DAF. (Photo: DAF Trucks Ltd)

Following good experiences with its previous 2800 tractor units, NC Cammack & Son continued to buy DAF. This smart DKS was bought new and generally worked on low loaders, which made up a lot of the company's work. However, when this side of the business was slow Cammack was quick to take on continental work, generally with tilts. Their 2800s seemed well suited to either role and performed both admirably. (Photo: Jim Cammack)

Among Fairclough's huge, multi-million pound itinerary of equipment were a number of dependable FT2800 DKTD units. Fairclough was heavily into civil engineering products and the DAF units worked on both own-account operations, moving the company's plant and machinery around the country, and the delivery of bridge sections. Once again it was the DAF reliability and back-up that convinced Fairclough to buy the trucks, as the penalties for late delivery were high in their line of work. *(Photo: DAF Trucks Ltd)*

UK general haulage was J&C Finney's speciality and in 1979 six FT2800 DKTD tractor units joined the family firm to take on such work. The company's first DAFs were bought in 1976 and impressed by returning good economy and reliability with high driver acceptance. With almost forty vehicles by 1979, many of which were FT2800 DKTDs, J&C Finney had come a remarkably long way from the modest start by brothers Jim and Cliff back in 1947. (Photo: DAF Trucks Ltd)

William Nicol's tankers were used to transport a number of liquid cargoes, including whisky. The rest of the trailer fleet at the time comprised forty-foot flats which were used for all manner of general cargo. The Aberdeen Company was established in 1936 and, by the time this FT2800 DKS was bought, had amassed an enormous amount of experience and a strong, loyal customer base. This photograph shows how well the three-wiper system cleared the large screen, especially in front of the driver.
(Photo: DAF Trucks Ltd)

Greenwoods Transport bought its first DAF in 1976 and after two years of running evaluation, which was company policy for all new types before further purchases, it bought more and more. This FT2800 DKTD was one of the latest batch of six identical units to join the fleet in 1980 and was the first DAF registered with the new W registration.
(Photo: DAF Trucks Ltd)

The benefits of the new 6 x 4 rear axle were not wasted on Arcade Motors which chose to base this handsome wrecker with Bulldog hydraulic equipment on the new FTT2816 heavy haulage tractor chassis. The 2698T axle featured six torque rods, which countered the forces of acceleration and – more importantly in this role – braking, giving better stability. The new unit also featured three electro-pneumatic diff-locks, one for each axle and one between them, very useful for recovery work. This was Arcade's third DAF truck. (Photo: DAF Trucks Ltd)

Exonia European might be better remembered for moving exotic boats, but this fine DKTD makes an equally impressive sight as it storms down the M4 with this neat and tidy load of cut timber some time in the mid-1980s when the unit would have been four or five years old. Note the wide axle spread on the trailer and the unit's vertical exhaust mounted on the nearside.
(Photo: Adrian Cypher)

The second-generation 2800s arrived in 1980. Key among the changes were those to the F241 cab, and the most significant part of that was the introduction of double-glazed windows for the rear. This answered the drivers' biggest, and long running, criticism of the otherwise superb working environment offered by the big cab. The glazing was only part of the story, though, for more insulation was fitted to the roof, walls and floor, and a new brown and beige colour scheme added to the new cosy feel. Many of the second generation improvements – certainly those relating to cab comfort – were a knock-on of the changes made when DAF started to market the 2800 in Sweden late in 1979. Note the new badging arrangement adopted at this time with model designation moved to the main grille opposite the DAF name.
(Photo: Dave Wakefield)

'Wirtschaftlichkeit durch Qualität' or 'Economy through Quality' was DAF's statement at the Frankfurt Show in 1979, and with that in mind they introduced a new economy model. The 2800 DKSE was all but indistinguishable from its DKTD and DKS stable-mates, save for the subtle 'E' incorporated into the badge. The DKSE engine was part of DAF's constant quest for better fuel economy. In theory, the DKSE fell between the DKTD and DKS by offering 276 bhp. However, the DKSE offered a substantial increase in torque over the DKS. DAF developed the DKSE as a low-speed, high-torque motor that would, in theory, have its peak power available at the most critical times and reduce gear changes and therefore use less fuel. The maximum speed was just 1800 rpm and the new engine was indeed very frugal, smashing many industry magazine test route records. But a narrow band of useful torque required a disciplined driver and the truck was slow in comparison tests of the time. (Photo: DAF Trucks Ltd)

WF Hall & Son's long association with DAF trucks started with SAD 299N, an original 2600 unit noted for its storming performance on the flat, but not the hills. Natural progression brought 2800s to the fleet with DDE 555W being the third. The DKS unit was the first in the fleet to feature the updated cab with the new cosy brown interior and was assigned to Martin Hall, son of the founder.
Safety improvements incorporated in the second-generation cab included a new design of self-draining cab step with a punched hole, rather than the previous grid pattern. The new steps were deeper, too, so they could be seen from above when descending. Cardboard and paper made up a lot of Hall's work and here another load is roped and sheeted prior to leaving Goodwick harbour. (Photo: Marcus Lester)

With the huge number of trucks travelling to the Middle East from Europe and the growing market for trucks in fast developing countries such as Dubai, DAF signed up an importer for the region in 1980. Saudi Arabian Truck Agencies (SATA), owned by HRH Prince Bander Bin Khaled Bin Abdul-Aziz, was based in Jeddah with satellite operations in Riyadh and Dammam. Staff were trained by DAF and the principal product was based on the heavy version of the European 2800, the 2805. Tractors and rigids were available and came supplied with modifications to suit the conditions, which included air tanks and batteries that were relocated to a position on top of the chassis behind the cab and air conditioning. This example is employed on brick and block work Saudi Arabian style. Note the height of the trailer bed.
(Photo: DAF Trucks Ltd)

This second-generation FT2800 DKSE continues Visbeen's association with DAF. Visbeen's first DAF truck, along with the rest of the small fleet, was lost to the surging waters of the devastating great flood of 1953. Happily, since then the North Sea and English Channel have taken on quite a different significance for Visbeen, acting as the conduit for up to 160 crossings a week around the time of this photograph. *(Photo: Adrian Cypher)*

Not the world's largest game of Jenga, but another impressive load of stacked wood destined for one or other of the many collieries serviced by RG Fowler in England and Wales back in the early 1980s. The South Wales haulier was among the very first to take delivery of a second-generation FT2800 with this DKTD-powered example, which was also its first DAF. Note the convenient positioning of the cab lift pump adjacent to the diesel tank. This was the right-hand drive chassis location for this item, keeping it safely placed on the nearside in breakdown situations; on a left-hand drive chassis it was generally mounted on the side – or underneath – the battery tray. *(Photo: DAF Trucks Ltd)*

Vales DAF Trucks was a main dealer from the start of DAF's UK operation and after just seven years supplied its 1,000th truck when it delivered this smart FT2800 DKS to George Fischer. The truck was expected to cover 70,000 miles a year delivering steel tube, iron and plastic fittings to customers throughout Europe and was well equipped for its role with a catwalk diesel tank, roof spoiler and this delightful tilt trailer. *(Photo: DAF Trucks Ltd)*

The time to tilt the big F241 cab was halved in a separate development announced early in 1981. The revisions included a new pump featuring a longer stroke with reduced bore together with a longer operating handle. Also, the old cab release valve was replaced with a new one mounted on the pump and was only operable by the same handle. As well as being quicker, the operation was now smoother too. *(Photo: DAF Trucks Ltd)*

McIntosh, the Aberdeen-based meat producer, was one of the first to put a DKSE-powered FT2800 to work on the meat run; an arduous, fully freighted and high-speed run from Scotland to the south of England. The truck was operated alongside two DKS units, giving McIntosh a superb opportunity to measure and compare the fuel economy of the new engine. While European-spec DKSE trucks were fitted with the ZF sixteen-speed gearbox, UK versions got a nine-speed Fuller unit, which allowed for a higher road speed of 66 mph. (Photo: DAF Trucks Ltd)

With significant road building contracts to supply that included two sections of the M25, CA Blackwell needed reliable prime movers for its extensive collection of bulldozers, excavators and dumper trucks. This FTT2816 was bought following the good experiences of other operators, particularly NC Cammack & Son which was based nearby and often subbed to Blackwell. The second-generation cab received a useful modification in the mirror department with the introduction of wider and heated units. The standard set-up was a five-mirror system with one on the driver's side and three opposite consisting of one normal view, one wide angle and one for the kerb, the latter being mounted at the top of the door. *(Photo: DAF Trucks Ltd)*

Sadly missed, but wonderful to remember that superb livery. Lloyds' smartly turned out fleet was constantly travelling into Europe. While all manufacturers offered breakdown assistance of one form or another, DAF could proudly say that during 1979 its ITS (International Truck Service) had managed to put 98 per cent of breakdowns back on the road within twenty-four hours. (Photo: Dave Wakefield)

This FTT2816 DKS of John Stacey shows what the 6 x 4 chassis was capable of at the top end of the scale. While able to operate at up to 150 tonnes under STGO 3, the FTT2816 was also suitable for more general low-loader work and, indeed, some operators even managed to utilise them successfully on general haulage when their main business was quiet. Note the bumper conversion with beefy towing eye. *(Photo: Dave Wakefield)*

It is easy to forget just how good the F241 cab was, considering it was among the first of the 'modern' cabs to appear in the 1970s and soldiered on for almost twenty years. Roomier and better packaged than an F88 and with a superior ride compared to an LB110, it was a driver's dream. Added to that, all daily checks were easily carried out through the main and lower grilles and items like headlight bulbs were quick to change with the entire cowl folding down following release with a coin. Dunkerley's plain but effective livery was always easy to spot. The company's 2800s were gainfully employed on block work and heavy haulage. *(Photo: Adrian Cypher)*

The big 1160 engine was fitted with a thermostatically controlled fan. This featured an electro-magnetic drive and it was possible to activate it manually from a dashboard switch if the driver detected a failure. It could also be activated prior to stopping to avoid heat soak. If both means of operating the fan failed it could be manually locked into permanent operation with a couple of bolts. All of which added to the truck's 'get-home' appeal and no doubt contributed to its popularity on jobs like the Middle East run. Proving that given a good truck a good driver will take pride in it, this handsome unit of Windebank Bros with its fine livery features many accessories of the day, including pressed wheel trims. *(Photo: Marcus Lester)*

It seems a long time since Carter's has pulled anything other than containers, but there was a time when flats like this were just as much a part of the business. Carter's had been running FT2800 tractors with the DKS engine for three years when it placed an order for six DKSE versions with high expectations for even better economy. DAF claimed the DKSE would better the DKS on fuel by 10 per cent and many operators corroborated that claim. Note the extra diesel tank on the offside of this unit, which must have afforded it a range of over 1,400 miles. (Photo: DAF Trucks Ltd)

An unfortunate experience with the F88 290 put Gallacher Bros off imported trucks, but when one domestic manufacturer was only able to deliver one of four trucks ordered on time, something had to give and an FT2800 fitted the bill. This opened the floodgates and by the mid-1980s the thirty-plus fleet was half-DAF. In addition to the superb reliability, Gallachers found that – on the rare occasions it was needed – the back-up service was second to none and the main reason for further purchases. If it were not for operating a mixed-fleet policy to keep manufacturers keen, DAF numbers may have been higher still for the Northumbrian-based company. (Photo: John Henderson)

Another smart Lloyds unit, though the trailer has seen better days. The official designation for a 6 x 2 tractor chassis was FTS2800. This unit, a DKS, could be as it left the factory or was possibly altered with a tag axle some time later to offer three-plus-three running at 38 tonnes. Many operators, especially those on container work, favoured this configuration as it safeguarded to some extent against axle overloads. However, the flip side was the increased weight of the unit and subsequent loss in payload. (Photo: Marcus Lester)

'Colne Valley Cavalier' was NC Cammack's latest fleet addition in 1982. This marvellous FTT2816 unit was generally kept busy moving large plant such as Caterpillar scrapers and the monstrous D9 bulldozers which tipped the scales at a hefty 75 tonnes. Despite a rather slow top speed of just 52 mph, the unit would also be used on general haulage when necessary and on one occasion did a trip to Italy with a tilt. *(Photo: Jim Cammack)*

In 1982 DAF announced the 3300. Although the new model was basically another re-tune of the existing 2800 and 1160 engine, DAF, for the first time, decided to give the vehicle a new designation to make the distinction between it and the DKTD-, DKS- and DKSE-equipped 2800s. As well as a '3300' grille badge, the new model was also given new badges that ran under the side windows. This change was also applied to the 2800 at the same time, as is well illustrated here by this smart FT2800 DKSE of Ings Transport. *(Photo: Marcus Lester)*

DAF offered a generous 12-month unlimited mileage warranty on all its trucks from new, such was its confidence in the product. In addition to that, owners could opt to pay a one-off extra charge based on the retail price of the truck, which would cover any unscheduled replacement of the power train components. On a DKTD-equipped tractor like this with a retail price of £25,550 in 1982, Power Train Protector, as the scheme was known, would have cost £500-£1,300 depending on the extent of coverage. DAF also offered a contract maintenance scheme on a fixed cost-per-mile basis. This usually ran for the first four years or 300,000 miles. (Photo: DAF Trucks Ltd)

This smart twin sleeper operated by Nippress prominently shows the new-style 2800 badges under the side windows. By 1980, DAF had made a policy change and the day cab was now the standard offering instead of the sleeper, so it now cost £1,300 if you wanted to kip in your 2800. But as always in haulage, haggling and deals were a large part of the buying process and list prices were more like guidelines for the salesmen. (Photo: Dave Wakefield)

Obviously spending much of its time on the continent, this UK-registered but left-hand drive FT2800 DKSE of Fransen UK is fully kitted out for economy with DAF's elaborate roof spoiler and under-bumper air dam. The DKSE was the first big DAF to be fitted with the VISAR driving aid. The system used a series of lights on the dashboard to advise on the best gear for any given situation and helped make the best of the narrow economical rev band of the DKSE.
(Photo: Dave Wakefield)

The conclusion of the May 1983 Armitage Report – named after Professor Armitage, the government advisor who did the study and came up with the proposals – was that the UK should accommodate a weight increase to 38 tonnes carried on five axles. Many operators favoured three-axle units for the new weight as the costs could be beneficial in terms of taxation and a three-axle tractor was more versatile as it could pull two- or three-axle trailers. This spurred the manufacturers to look at new ways of accommodating this move. The twin, or rear-steer, chassis suddenly found itself back in vogue for the UK after an absence of many years. This smart DKSE of Bulwark shows DAF's neat and tidy solution, designated the FTG2800 (G – a 6 x 2 twin steer), allowing the operator to run legally with this two-axle tanker. (Photo: Marcus Lester)

Although day versions of the F241 cab were in the minority, Robson's of Carlisle did operate a great number in the famous maroon and cream livery of its intensive trunking fleet and not until the mid 1980s, as operational requirements dictated, did sleeper versions start to appear. All Robson trucks bore an individual name, often chosen with the driver's input and were generally matched to a dedicated trailer. *(Photo: John Henderson)*

A plain and simple livery marks out this smart combination of an FT2800 DKSE and tri-axle gas tanker. The introduction of the 3300 in 1982 made the old DKS-powered 2800 somewhat redundant and the model was dropped from the UK range. A standard-spec DKSE tractor was plated at 41 tonnes GCW and was thus capable of most duties and, being around £1,000 cheaper at standard list price than the FT3300, found a useful niche in the market. Yorkshire-based LPG also operated 2300s and 2500s and ran a slight bias towards two-plus-three configurations. *(Photo: Marcus Lester)*

This extremely smart FT2800 DKSE with just 276 bhp was very much at the entry level for West of Scotland's impressive fleet. However, the useful low-down torque suited the company's purpose well and despite being a 4 x 2 was perfect for getting loads like this on the move. The Scammell S26 was a cancelled order which was originally destined for Africa and was bought at short notice for a good price. The unit was left-hand drive with a Cummins/Fuller drive train. Plated to 100 tonnes it performed well until the chassis finally broke. *(Photo: Jimmy Campbell)*

Another good example of the twin-steer arrangement – and this time on a DKSE tractor fitted with the full petrol-regulation package, including front-mounted exhaust and steel battery cover. The DAF system, unlike some competitors, featured a positively steered second axle rather than a self-tracking item. This obviously added weight, but also meant a raised profile for the cab, as everything was slightly higher to allow room for the steering linkage. *(Photo: Marcus Lester)*

These four FT2800 DKTD tractors joined the Hargrave fleet at the end of 1984, making them the last of twelve DAF trucks bought by the Spalding-based company that year. With an intense refrigerated operation, Hargrave ran its DAFs hard, many, including these, were double shifted and averaged more than 130,000 miles per annum. Reliability was therefore of paramount importance and Hargrave did not find the DAF product wanting. (Photo: DAF Trucks Ltd)

In 1984, after just 35 years as a commercial vehicle manufacturer, DAF produced its 250,000th truck. Rather appropriately, given its huge importance to the company, it was an FT2800 tractor, though not this one, that claimed the honour. The year also marked significant changes to the Eindhoven plant with an investment of £¾ million that radicalised the production line, increasing capacity to eighty vehicles a day. (Photo: Marcus Lester)

B30 JSS was the first twin-steer to join the Bon-Accord fleet for its refrigerated operation which covered the entire UK. The FTG2800 DKSE unit was plated to 46 tonnes as standard and came equipped for economy on those long journeys south with DAF's roof spoiler and under-bumper air dam, both of which provided their best results at motorway speeds. Note the neat revision to the lower part of the battery box that appeared around this time, but only on the twin-steer. *(Photos: DAF Trucks Ltd)*

This FT2800 DKSE took NETransport's fleet to twenty-five trucks. The Montrose company enjoyed quick growth servicing the agricultural sector with a fleet of three-axle bulk tipping trailers. As such this unit, plated to 46 tonnes, was equipped with hydraulic tipping gear and tank. When NET once suffered a complete engine failure in Carlisle, the Marlow headquarters supplied a new one and the truck was on the road again by the following afternoon. It was this kind of back-up that kept the fleet 75 per cent DAF. *(Photo: DAF Trucks Ltd)*

The first DAF joined the LPG fleet in 1976 and just eight years later the company bought its 100th. Part of an order for five FT2800 DKSE units, the new truck, with full petrol-regulation spec, was put to work for the company's top clients, which included BP, Esso, Procter & Gamble and the US Airforce. LPG monitored their DKSE-powered trucks closely and noted a good average consumption of 7.25 mpg at 38 tonnes. *(Photo: DAF Trucks Ltd)*

Although the 2800 was still an excellent product in 1984, many of its key competitors were now much newer designs. While DAF was well advanced in the development of its own replacement vehicle range, this was still several years away and something was needed now to inject new life into the old 2800. The first stage of this final revamp was the introduction of the Space Cab. With a towering glass-fibre roof extension, DAF created a super premium truck to cater for the needs of long-distance drivers. *(Photo: Dave Wakefield)*

No shrinking violets, the bright red fleet of Lincolnshire-based Denby Transport. Despite totally eradicating the manufacturer's name from its trucks, Denby has been a loyal DAF operator since the original 2600 and besides the 2800, also ran 2300 and 2500 examples. It was also an early taker for the new 95 series when it replaced the old 2800, 3300 and 3600. Note the long wheelbase of this example, probably the longer 3.5 metre option. (Photo: Dave Wakefield)

By 1985 when J&C Finney was operating from a six-acre site with 20,000 square feet of warehousing and paper, oil and coiled steel movements had become something of a speciality. That year the firm also bought its 100th vehicle, this smart FT2800 DKSE. The new unit joined six other DKSE tractors and operated with a mix of trailers including three-axle flats at 38 tonnes. Finney's renewal policy saw trucks replaced usually every two years and residuals were reported to be very good on the big DAFs at that age. *(Photo: DAF Trucks Ltd)*

Northern Ireland was an important market for DAF and a main dealership was set up in Ulster way back in 1978. Northern Ireland Carriers (NIC) was just one of the customers that they were keen to attract. With major customers such as ICI and SPD, NIC needed reliable trucks and this FT2800 DKTD was added to the fleet in 1985 to fulfil such a role. Many of NIC's trucks ran in customer liveries. *(Photo: DAF Trucks Ltd)*

After a year of 38 tonne operation in the UK, DAF confirmed that it still recommended a two-plus-three arrangement for UK operating conditions. However, it still offered the FTG2800 for those who wanted to run three-axle units and container operators definitely fell into that camp. With ISOs often badly loaded and front-end heavy, the extra axle on the unit helped reduce the risk of axle overloads. The twin-steer arrangement also offered good manoeuvrability and better tyre wear than other 6 x 2 layouts and improved ride and stability over a 4 x 2. Probably among the last of the old DKSE-powered 2800s, this well-worked example looks very purposeful while at rest. *(Photo: Marcus Lester)*

Advanced Turbo Intercooling or ATi was to be the second part of the final revamp for the old 2800 range. The ATi re-brand also covered the 3300 and launched a brand-new model, the 3600. In essence the ATi engines were early releases of the next generation developments intended for the forthcoming 95 series trucks with the Cabtec cab. Under the new branding a 2800 truck got the DKXE version of the old 1160 engine offering 288 bhp and 959 lb/ft of torque; this was to be the new super economy version replacing the previous DKSE. However, for those who wanted it, the old 253 bhp DKTD was still listed in the UK without the ATi branding. Note the new lower grille, which was also introduced at this time. (Photo: DAF Trucks Ltd)

The best of all the final developments here wrapped into one convenient package, an FTG2800 DKXE operated by the well-known Scottish haulier Davidson & Wilson. Unlike previous UK versions of the DKSE, which had to utilise the nine-speed Fuller gearbox to attain UK motorway speeds, the new DKXE was fitted with the sixteen-speed ZF unit that European DKSEs had been fitted with since 1979. This was thanks to the new rear axle fitted to the DKXE (the 1346 single-reduction type), which came with a 3.31:1 ratio diff as standard. (Photo: Dave Wakefield)

Another example of the complete package, but this time a handsome FT2800 4 x 2 unit operated by Colchester-based Willsher Transport. As all the glass behind the driver was deleted in the Space Cab it was no longer necessary to dog-leg the air stack as before, so all examples, left- or right-hand drive, utilised the same straight version with extra length to stretch over the high roof extension. Note the additional diesel tank mounted on the offside, which has necessitated a single battery and air tank per side as a revision. (Photo: Marcus Lester)

RABA, the Hungarian truck builder, first started to build trucks with DAF cabs and side frames in 1980. By 1985 it had produced more than 3,300 such vehicles and a new contract was signed with DAF to continue the supply for another five years. Although the cabs were outwardly the same, they were supplied without seats and bunks, RABA fitting less luxurious items. Engines were licence-built from MAN and a good deal of production went to the State-owned Hungarocamion. The trucks, though simple and robust, were expensive compared with imported brands like Mercedes because of the purchase and shipping of major components like the cab. However, RABA enjoyed good export markets in North Africa, the former Yugoslavia and even China for its high-bred trucks. (Photo: Adrian Cypher)

In fine fettle at one year old, this FTG2800 ATi of Gibb's of Fraserburgh looks simply stunning in the company's fine livery matched with this fridge trailer by Crane Fruehauf. The trailer was one of the last of this type used by Gibb's as new replacements from neighbours, Gray & Adam's, were steadily becoming the norm in the fleet. The unit, named 'The Scottish Soldier' is seen parked in Tower Street, Leith by the loading bay of the North British Cold Store. (Photo: John Henderson)

Given that all the big DAFs were generally noted for excellent fuel economy, it is a little surprising that Gibb's found the 3300 wanting in that department compared to its established ERF fleet. However, by down-sizing to the 2800 the situation was rectified and the company subsequently bought fourteen FTG examples to operate at 38 tonnes as three-plus-three combinations. As economy was high on the list for this long-distance fleet, all the Dutch machines were fitted with roof spoilers and under-bumper air dams. *(Photo: John Henderson)*

Apart from a rusty, pockmarked bumper, this FT2800 operated by the shipping company Nedlloyd, looks in great condition for a truck with over ten years work under its wheels. Although the F241 cab was not immune to the dreaded tin worm, it did seem to fare much better than many of its contemporaries, including those from Scandinavia, and the later units were particularly resilient thanks to huge investments in the cab production facility at Westerloo. *(Photo: John Henderson)*

Fransen Transport first ran outfits like this in late 1986. The combination could carry two extra pallets, 26 in total, as the reduction in the BBC (bumper to back of cab) measurement of the top sleeper equipped cab allowed for a trailer with a 13.1 m long deck, while remaining within the 15.5 m limit. To aid close coupling, the trailer was fitted with an under-slung refrigeration unit by Petter. DAF first started to build top sleeper cabs as far back as 1980 to satisfy the requirements of operators with loads that would frequently 'cube out' and was commonly fitted to continental drawbar outfits. The sleeping compartment was accessed via a lockable fold-down hatch and step. For safety there was a large emergency exit fitted in the roof, which doubled as skylight/vent. This fine post-ATi FTG2800 made a purposeful three-plus-three combination and was employed on international runs with loads of fresh and frozen vegetables, pastries and ready meals. *(Photo: Marcus Lester)*

Perhaps DAF's ultimate development in maximum load carrying was the somewhat ungainly MAG2800. Co-developed with DAF's special projects partner, Ginaf, the unique machine had the capacity to carry five air-cargo containers, three in the main truck and two on the close-coupled trailer. The truck's chassis was 6 x 2 and featured a second steered axle; one of the trailer axles also steered giving good manoeuvrability. The modified cab was even shorter than the normal top sleeper thanks to the deletion of the small rear quarter-window and was fixed to the custom chassis. The engine was a horizontal version of the 1160 originally developed for use in DAF coaches. Ginaf's relationship with DAF was much like that enjoyed by Terberg and Volvo and often produced similar multi-drive results for the off-road market. (Photo: Dave Wakefield)

This FTG2800 is something of a conundrum, wearing the badging of a DKS (turbocharged and intercooled), but registered after the introduction of the DKXE ATi when the former model was no longer available. A special order, perhaps, or maybe a later registration of an older model, although, if the latter, then the truck was updated with the new lower grille. Whatever the answer, it made a great-looking truck. (Photo: Marcus Lester)

Hallett Silbermann ran a large number of DAF 2800 tractors with most configurations accounted for including 4 x 2, 6 x 2, 6 x 4 and 6 x 2 twin steers as illustrated here. With the company's varied operation, which could be anything from low-loader work to continental tilts, this gave it great flexibility. This interesting load is carried on one of the company's standard three-axle flats. Note the large, non-standard mirror on the offside. (Photo: Marcus Lester)

Not as precarious as it first appears, this load of new muck-spreaders is actually securely located to the trailer via custom-made transport frames. Nonetheless, it makes for an impressive sight, especially with one of Evans and Williams' superbly turned out FTG2800 units heading it up. With the first of the new 95 series DAFs also appearing on E-registrations, this unit marks the end of a fantastic period for DAF and the 2800.
(Photo: Marcus Lester)

DAF 3300
What's in a Name?

By 1982 the big 2800 DAF had been around for nine years and established itself as a top-class truck with an enviable reputation for economy and reliability. On the occasions when reliability failed, DAF was not found wanting with its excellent DAFaid system always coming to the rescue.

However, things do not stand still long in the world of heavy truck development and since the introduction of the 2800, many other manufacturers had launched new ranges and the power rating for super premium trucks had climbed steadily upwards. Enter the new 3300.

Solidly based on the 2800, which would still be produced alongside the new truck, the 3300 used the same chassis and running gear as before. The cab was the familiar F241, complete with the recent second-generation updates. A new '3300' grille badge marked out the new model, as well as additional badges which ran below the door glass and under the side window of the sleeper. The badging was perhaps more significant than it appeared as it was the first time that DAF had treated an engine development of the 2800 as a separate model, giving it an identity all of its own rather than a set of letters denoting an engine option. Obviously, this was driven by the marketing department, but it was definitely the right thing to do.

At a stroke, DAF had apparently produced a new model that, with bigger and better numbers attached, placed itself nicely amongst the other big hitters out there. While not on a par with Scania's 142, the intercooled F12 from Volvo or the Mercedes-Benz 1948, it did measure up well to trucks like the standard F12, Renault's R340 and a host of others, like the Transcontinental, which were available with the big-power Cummins engines.

DAF had already produced a 350 bhp (DKSI) version of the familiar six-cylinder 1160 engine for the Italian market, where high weights, mountainous terrain and a requirement for 8 bhp per tonne dictated such outputs. Furthermore, as DAF's engine test facility had run examples without trouble at up to 400 bhp, the engineers were confident of maintaining the legendary DAF reliability in a unit producing 330 bhp.

The new engine was designated the DKX 1160 and, like the DKS and DKSE, it was turbocharged and intercooled. The maximum operating speed was 2200 rpm – the same as the DKS – and this coupled with a new KKK turbocharger and a matched Bosch fuel pump did much to help the engine achieve the new output. Although the increase in brake horsepower over the DKS was fairly modest, the torque and its delivery were very different, and indeed much improved over both the DKS and DKSE versions. Over 800 lb/ft of torque was available between 1000 and 1800 rpm with a very useful maximum of 966 lb/ft generated at just 1300 rpm. Not only was that over 100lb/ft more than the DKS, it was delivered at 150 rpm less too. And while the DKSE could muster a creditable 929 lb/ft at the same revs, its narrow window of availability required great skill from the driver if good economy was to be achieved.

However, with that sort of torque available it was possible to get carried away, so DAF installed the VISAR driving system as standard with its three warning lights and graded dial prominently placed in front of the driver. The gearbox was the double-H pattern Ecosplit unit from ZF offering 16 speeds, which was combined with the DAF-manufactured 2699 hub reduction axle fitted with the 4.49:1 ratio diff. This gave a sprightly performance and, with nearly 32 mph per 1000 rpm in top, meant a giddy 71 mph maximum but, more importantly, a very relaxed and economical, legal cruise. For those who habitually operated in mountainous terrains, an optional final drive with a 5.03:1 ratio gave improved gradeability at the cost of a lower top speed and less economical cruise.

Two years into its production life, DAF, keen to revitalise the existing range and keep it fresh until the new models scheduled for 1987, introduced the Space

Cab. Through market research, DAF learned that besides a comfortable cab with plenty of storage space, one of the most important features for the long-haul driver was the ability to stand up in the cab for dressing etc. DAF's solution was a truly striking one that saw the F241 cab grow in height by nearly two feet. Following the same rake angle as the windscreen, the fibreglass addition was big and boxy in design to maximise the extra space inside and, on tractor units, was mated to aerodynamic screens between the back of the cab and trailer which did much to balance the extra height when viewed in profile.

Inside the twin bunks, with safety nets, were relocated creating a useful under-bunk storage compartment below the bottom one and allowing the occupants of both more headroom. At the end of the lower bunk was the usual space for a toolbox, but a purpose-made TV and video unit mounted on a special turntable could be specified to sit above this area. Above the screen was a shaped console containing two lockers, an open space for oddments and all the electrical items such as the radio/cassette and CB. The shape of this unit left the area above and in front of the passenger seat free, allowing a six-foot-plus person to stand upright.

Insulation of all F241 cabs was increased at this time, with the 'artic pack' usually reserved for Scandinavian countries becoming standard fitment, and in the interest of comfort, all rear glazing was also deleted. The floor, engine cover and EDC (Electrical Distribution Centre) were all covered in a warm, brown carpet, and fully adjustable heated seats with wool covers could also be specified. Big and boxy, the Space Cab made a bold statement and found instant favour with any driver lucky enough to be assigned to one. Fleet flagship, owner-driver's statement or pure indulgence, the Space Cab was a remarkable success.

This smart FT3300 with matching van trailer would have been among the first examples registered in the UK and wears the X-plate applicable for 1982. The 3300 was essentially another engine development of the 2800 but, for the first time since its launch in 1973, the truck was designated with a new number which created a separate model from the 2800. New 3300 badges were prominently mounted on the grille and on black plastic strips below the side windows. *(Photo: Marcus Lester)*

With such strong loyalties to DAF, it was no surprise to see Visbeen running some of the earliest examples of the new 3300 in the UK. This X-registered example, an FT3300, was among a batch of eight such units on the UK fleet which undertook the distribution of the goods from the company's warehouses in England to wholesalers throughout the country – at the time around 6,000 trailer movements a year. Visbeen's 150th DAF truck purchase was an FA3300, high-capacity drawbar combination for the Dutch-based fleet.
(Photo: DAF Trucks Ltd)

The 3300 was launched to an eager industry in 1982. The trade press were quick to test the new model and extracted remarkable economy, always a trademark of the 2800 – particularly the DKSE, while returning good journey times. Part of the credit for this frugality had to go to the onboard Visar driving device, which advised on gear selection, but the superb engine mapping had a big part to play too. Ultimate mpg was not necessarily the highest priority for British Ropes with this heavy-duty FTT3300 example. Far more important in this application was the 966 lb/ft of torque available at 1300 rpm. Note the repositioned air tank ahead of the second axle. (Photo: Dave Wakefield)

'Hardhill Cavalier', an early FT3300, was the only one of its type to sport D&J Sibbald's smart traditional livery and was employed on general haulage duties throughout the UK. With no allegiance to any one manufacturer, Sibbald's fleet was an eclectic mix with most manufacturers represented at one time or another. The Bathgate trucks were named after the company's base at Hardhill Farm. (Photo: John Henderson)

IPEC's remarkable thrust into Europe was achieved with an open chequebook in the late 1970s. By 1982-3 IPEC's European fleet numbered many hundreds of vehicles. Of these a high percentage were the final-stage delivery vans and light trucks, but a fleet of seventy 2800 and 3300 drawbars and artics were among the company's heavy fleet, which carried out the trunking duties between hubs. The drawbar spec was the same for both 2800 and 3300 trucks and featured this 6 x 2 (FAS) chassis layout with a trailing single-wheel tag axle. Trailers were a conventional two-axle type and the box or tilt bodies carried were de-mountable to minimise turnaround times. One of the significant companies bought by IPEC to start the European business was a Dutch company, Gelders-Spectra, which could explain the high dependency on DAF products, although there was also a large DAF dealership next to IPEC's Arnhem depot. Rivals TNT took over IPEC's European activities in 1983. (Photo: Dave Wakefield)

An atmospheric early morning start for TIA 1633 makes an evocative sight as it sweeps, power on, through a tight left-hand bend. G Curran & Sons operated what was believed to be the first 3300 registered to an Irish operator, which followed a number of 2800s into the company's old red and white livery. Perhaps the registration of this unit would have better suited to a contemporary Mercedes, a marque that would later come to dominance in the Curran fleet. *(Photo: Dick Curran)*

WF Hall & Son's general haulage business enjoyed steady growth through the introduction of a new ferry service between nearby Goodwick harbour and Rosslare in the early 1970s. With the help of an expanding fleet of FT2800 DAFs, the company tailored a service for goods in and out of Ireland and the UK which continues to this day. This smart FT3300 was an early example of the model and was the personal truck of Martin Hall, who took over the running of the company from his father in 2002. In 2005, Hall's livery was changed to a smart blue and white scheme. *(Photo: Marcus Lester)*

This fine example of an FTT3300 operated by Stiller of Middleton St George looks extremely smart in the famous yellow and blue livery and shows what a perfect heavy haulage tool the big DAF was with its high cab, plenty of glass and excellent mirrors to allow the crew to monitor extreme loads such as this. Note the second Stiller unit, a 2800, lurking behind. (Photo: Dave Wakefield)

The FTS3300 (6 x 2) made an ideal tool for TIR. The single-drive 2699S bogie with air-actuated diff lock had a capacity of 20,000 kg, giving the FTS chassis a design capability of 55 tonnes, ample margin for those operating at 38 tonnes. The new DKX engine, like the old DKS, ran to a 2200 rpm maximum, but delivered its 966 lb/ft of torque at a lowly 1300 rpm. This gave the 3300 a great amount of get-up-and-go, which enabled traffic speeds to be attained quickly so that good journey times could be achieved even in urban environments. (Photo: Marcus Lester)

Merzario road/rail tilts were a common sight behind BJ Myers' trucks as the company specialised in pulling continental trailers throughout the UK. The Woolwich-based firm was a staunch DAF user, running over twenty examples of the 2800 and 3300 in a fleet that included Mercs, Volvos, Scanias, MANs and Leylands. This example features a Granning lift axle and was one of two such units bought to facilitate three-plus-two running before DAF's twin-steer chassis became available. Later, large numbers of Iveco 220.30 twin-steers were purchased and, while the lower weight of the Italian machines gave greater payload, it was found that brake, clutch and tyre lives were poor in comparison to that of the DAFs. *(Photo: Marcus Lester)*

Sunshine on Leith. Pollock driver Peter Wight puts his feet up and takes ten in the shadow of a massive dock crane while waiting for his load to surface at the docks. Pollock did good business transporting mini submarines for Vickers Oceanic, and understood the finer points of moving these expensive machines without damaging them. Note the Michelin man perched on the mirror arm of 'Invincible' and the catwalk diesel tank fitted to improve the range of the truck on those long hauls south to London which made up so much of the company's work. (Photo: John Henderson)

The merging of Wall's and Birds Eye in 1982 created the largest producer of ice cream and frozen food in the UK. The resulting change in operations saw vehicle mileage increase by 350 miles a week to 1,200. As part of a major fleet investment programme, twenty-nine FT3300 tractor units were bought, marking a policy shift away from the previous supplier, Scania. Although only running at the old 32 tonnes limit, the trucks were fully loaded for 75 per cent of the time and the extra power of the 3300 was deemed desirable to reduce maintenance in the long term. Note the under-slung chiller arrangement of the trailer. (Photo: DAF Trucks Ltd)

This interesting comparison between general and heavy haulage duties being performed by the same tractor unit shows the versatility of the 3300, especially as a 6 x 2. Operators that could afford the extra weight of this configuration on general haulage had the perfect multi-role truck capable of moving impressive loads such as the enormous Lancer Boss forklift. The only concession made by the operator was the addition of warning beacons and the compulsory signage. Interestingly, this unit was fitted with the heavy-duty Cyclone air stack and filter system usually only found on trucks which worked in extremely dusty environments.
(Photos: Dave Wakefield and Peiter Kroon)

Although DAF championed the two-plus-three solution for 38 tonnes operation in the UK, it was, of course, obliged to produce a twin-steer chassis too. The wheelbase was a reasonable 3.8 metres and both rear axles were air suspended. The repositioning of air tanks and ancillaries was neat, but produced a very busy chassis for fitters.
(Photo: Adrian Cypher)

The FTT3300 had a design Gross Combination Weight (GCW) of 56-125 tonnes and utilised the 2699T double-drive rear axle assembly in a 3.6 metre wheelbase chassis. The 330 bhp of the DKX engine gave a useful increase over previous heavy versions of the 2800 and, all in all, gave operators a good standard product for heavy haulage for around £35,000. This fine example operated by Allelys looks on top of this locomotive load of around 100 tonnes.
(Photo: Adrian Cypher)

DAF was the only significant foreign brand represented in the Carman Transport fleet in the early 1980s and made the numbers up beside products from domestic rivals, ERF and Foden. The Scholar Green based company had its roots back in the 1930s when William Carman started to haul stone for the local council. The company was always at the forefront of developments, being among the first UK firms to venture into Europe and the Middle East. This FTG3300 twin-steer makes a smart three-plus-three combination as it heads out with another temperature-controlled load destined for the continent. *(Photo: Dave Wakefield)*

International Transport was the first Scottish haulier to operate a 3300 tractor with the new FTG twin-steer chassis. The Perth-based unit was used on continental deliveries to fifteen countries with loads ranging from fish fingers to meat, and clocked around 70,000 miles a year. The truck was fitted with air suspension and ran at a full 38 tonnes for the majority of the time. With high-value frozen loads, DAF's back-up was a major part of the buying decision. *(Photo: DAF Trucks Ltd)*

Chris Tucker was one of the intrepid few to risk his livelihood by racing his working truck during the early years of the sport in the UK. The FT3300 DAF was the flagship of his Essex-based four-truck operation and took to the track largely unmodified, although, after a few such outings the big DAF was replaced by a much lighter, and expendable, day cab equipped ERF. During its racing career the truck was fitted with side skirts and ran without the massive DAF roof spoiler. *(Photo: Marcus Lester)*

Another firm that found the 3300 more than capable of the long haul down from Scotland and out through Dover on to the continent was G&J Jack. The family firm still specialises in the movement of fresh seafood from its Fraserburgh base, and the fish market at Boulogne is still a common destination. Premium tractors have always been the order of the day and are relied upon to deliver loads while in tiptop condition. *(Photo: Dave Wakefield)*

Although DAF had long offered a 6 x 2 tag axle chassis for tractors (FTS) it was not officially listed as available to the right-hand drive UK market until the introduction of the FTS3300. The wheelbase was 2.97 metres with a further 1.34 metres between the bogie. This example makes a good-looking combination coupled to this splendid European tilt. *(Photo: Adrian Cypher)*

The irony of this photo, showing a Nuttall FTT3300 just off the boat at Dover, is that the firm was part of a partnership of British companies that were awarded the phase 2 works for the Channel Tunnel. Nuttal was a tunnelling expert, having worked on those under the Mersey, Tyne and Thames at Dartford. The knowledge required for civil engineering underground was perhaps learned during the war years when the company built underground magazines, amongst other things, for the military. Nuttall was acquired by the Royal BAM Group in 2002, becoming BAM Nuttall, and is currently busy building for the 2012 Olympics. Own-account movements of the substantial amount of plant that the company needed kept this sturdy 6 x 4 unit busy. Note the catwalk tank, maybe to assist range in these Channel dashes. *(Photo: Dave Wakefield)*

The Charles Alexander fleet trucks were no strangers to this road, the A1, and frequently plied up and down it with loads of fresh fish destined for London. The company livery, a simple blue and white affair, sadly disappeared in the early 1980s and was maybe the first warning sign of the demise of this once great fleet. Day-cab versions of the 3300, much like the 2800, were definitely in the minority but some operators did find occasion to utilise the type, generally to accommodate the maximum length trailer possible. *(Photo: John Henderson)*

According to its livery this was not the usual role of this fine left-hand drive FTS3300 of Rowlinson Plant. Indeed, the truck's specification definitely leans towards a heavy haulage application and includes the fitment of the high-capacity cyclone air stack. Mostly fitted to desert-bound trucks, this item did find its way onto some UK tractors too and featured a straight cylindrical body mounted onto the twin cyclonic filter system. *(Photo: Marcus Lester)*

Robin East established Rokold as an owner-driver with just one FT2800 unit, NRO 842P. Performance and, above all, reliability were considered first class and more and more FT2800 units were bought as the business expanded. After 1982, buying also included the new FT3300. DAF's stronghold on the fleet continued until a change in operating policy required twin-steer tractors, at which point DAF's product was deemed too heavy.
(Photo: Dave Wakefield)

To enable clearance of the linkages required to steer the second axle of the FTG chassis, DAF's engineers had to raise the height of the big F241 cab. Although the difference is subtle, it is most obvious by comparing the height of the front step to the hub centre of the front axle with that of a 4 x 2 example. This raising also gave the impression that the cab of the twin-steer was leaning back on its rear mounts, especially when travelling at speed on a motorway. (Photo: Marcus Lester)

The crew of this well worked FTT3300 of Hills Heavy Haulage grabs a rest during proceedings. By the time of the launch of the 3300 model in 1982, DAF was confidently offering a two-year guarantee on cab paintwork and a five-year guarantee on corrosion in the panels, but the only colour option was now gloss white and the vehicle had to be re-treated by wax injection at one year old to qualify. This example had a substantial catwalk diesel tank to supplement the standard 400 litre item and was fitted with an upright exhaust on the nearside. (Photo: Marcus Lester)

A first-class example of Boom's immaculately turned out fleet arrives on UK soil in the late 1980s. This combination with top-sleeper equipped FT3300 nicely illustrates the maximum length combinations that were so popular with Dutch operators at the time. DAF was a big player in this market, which was hotly contested with rival products from MAN, Scania, Volvo and Mercedes-Benz. Note the self-tracking rear axle on the trailer. (Photo: Dave Wakefield)

A fine left-hand drive FTS3300 demonstrating the advantage of three-plus-two running as it bowls along with this tilt trailer in tow. The hub reduction axle of the FTS chassis could be fitted with either a 4.49:1 or a 5.03:1 final drive, giving 71 mph and 63 mph respectively. Trucks that spent the majority of the time on UK work would often run with the faster option, but those which were frequently on the continent, which had a lower speed limit at the time, may well have benefited from the improved hill climbing offered by the 5.03:1 option. (Photo: Marcus Lester)

The disguise is not fooling anyone here as this impressive FTT3300 and four-axle low loader combination undertakes the movement of a facsimile of the Reverend Audrey's most famous character. Apart from the 6 x 4 bogie, the FTT chassis differed from others in the range in being fitted with different-size wheels, a heavy-duty subframe and spring dampers on the front axle, the latter with the intention of countering nodding under load. Not too surprisingly, the Visar driving aid was generally deleted from the FTT. *(Photo: Marcus Lester)*

Despite the availability of the Supercontinental, which was equipped with many home comforts for the long haul, and then the Space Cab with its cavernous interior, the standard F241 sleeper was still found to offer a comfortable and spacious environment for the time and was very popular with drivers. Indeed, large numbers of 2800 and 3300 tractors with the standard cab, like this one operated by Falcongate, made gruelling trips to the Middle East, sometimes double-manned. *(Photo: Adrian Cypher)*

H Baker had been DAF customers since 1973 and, when this smart FT3300 joined the fleet in 1984, the original FT2800 unit was still working with half a million miles on the clock. The N-registered truck was frequently included in 2800 brochures to represent a typical UK-operated example and, with Baker's superb and traditional red, green and white livery, often outshone the more bland foreign examples alongside it. In 1984, the FT3300 was one of nine DAFs employed on UK general haulage duties for Baker's. (Photo: DAF Trucks Ltd)

Sutherlands ran around forty-five DAFs at the time this FT3300 joined the fleet in 1984, including a six-year-old FT2800 fitted with the DKTD engine, the company's first DAF. B672 GSA was one of two FT3300 units bought for a contract with British Caledonian Airways, which involved an air-freight trunking service between Prestwick and Gatwick airports. Sutherlands operated 130 trucks at the time and was the largest supplier of transport to the Orkneys and Shetland. *(Photo: DAF Trucks Ltd)*

Pollock's one-driver, one-truck policy generated great pride in the machinery. Phil Young, pilot of the immaculate 'Discovery', was no exception and kept this unit spotless. With the steady changeover to curtainsiders, it is perhaps surprising that Pollock did not fit roof spoilers to its tractors, especially given the trunking nature of its operation. B30 LSF was, however, fitted with the under-bumper air dam which did much to smooth the airflow under the truck. Note the neat installation of spray-suppression equipment inside the front wheel arch. (Photo: John Henderson)

The Dutch influence is clear to see on this FA3300 drawbar combination operated by Spalding Haulage. Indeed, the top-sleeper truck was fitted with an underslung refrigeration unit of Dutch design, and the centre axle drawbar was manufactured by Lamboo on a Groenewegen chassis, both Dutch companies. Spalding Haulage also operated a pair of FTG3300 tractor units and delivered plants and flowers to 150 outlets throughout the UK. (Photo: DAF Trucks Ltd)

Salford Van Hire's spot-hire fleet numbered 1,700 commercial vehicles in 1984 with a further 600 available for contract hire. Considering that the fleet included everything from vans to 38-tonners, DAF enjoyed good representation with around 200 trucks in the 16 tonne plus categories. B522 KGU, an FT3300, was new to Salford in 1984 and was part of an order which also included an FT2800 tractor and an FA2100 rigid. The 3300 was used for distribution work at 32 tonnes with this twin-axle reefer by Unger Meats of Manchester. (Photo: DAF Trucks Ltd)

In late 1984, B255 CCA became the latest step in the changeover to a 100 per cent DAF fleet for Cropper's Prepacks Limited. The company's work included the early morning collection of produce from farms and market gardens, which, with the nature of fresh food, demanded great reliability from vehicles. The firm had been running three FT2800 units since 1981 and, with the exception of a water pump and a shock absorber, these had completed a combined one million miles with just standard maintenance. Fitted with DAF's full air kit, the FT3300 operated at 32 tonnes and was confidently expected to return 9 mpg.

(Photo: DAF Trucks Ltd)

When a McIntosh driver was stranded at Sandbach services with a broken trailer spring and called DAFaid, the service dealer at Deeside cannibalised one of its own trailers to get the stricken truck back on the road. With back-up like that, and next-day loads that could exceed the value of a new tractor unit like this FTG3300 by £10,000, it is no wonder that McIntosh standardised on a 100 per cent DAF fleet for its meat wholesale distribution business. This FTG3300 was bought to pull the step-frame cattle trailer, which was used to supply the company's plant with livestock from surrounding farms. The Craven Tasker chassis had bodywork by EC Paton of Netherley and could accommodate forty cattle.

(Photo: DAF Trucks Ltd)

Designed to provide the best possible accommodation for the long-haul driver, the Space Cab made the ideal tool for those running to the Middle East, and B&T Hicks of Newport was the first to put one on the road in Wales. Hicks would despatch four or five trucks a week – half the fleet – to such distant destinations as Oman, and the Space Cab joined eight standard-cab 3300s on this work pulling a mixture of tilts and fridges loaded with general cargo.
(Photos: DAF Trucks Ltd)

By March 1985 EST's first DAF, EDP 774V, had clocked around 200,000 miles without a major problem. This was the kind of reliability EST liked and, having been wowed by the publicity material for the new Space Cab, an order was placed, sight unseen, for a new FT3300 equipped with the new high-roof cab. B377 BPU was the first Space Cab to enter service in the UK and was entrusted to long-serving driver Ted Winfield for its inaugural job supporting Joan Armatrading's '85 tour. Two more units (B378 BPU & B379BPU), also supplied by Harris Commercials, soon followed. EST also took the opportunity to update the livery at this point with a move away from the swooping stripes worn by the company's F12s to this more angular design. *(Photo: Dave Wakefield)*

Craib started out in 1973 with a contract delivering bank paper to London, but grew through its involvement with transport for the oil industry in Aberdeen. A merger with Aberdeen Road Runners in 1982 was beneficial to both companies. This FT3300 unit joined fifteen other DAFs, mainly FT2800 DKTD and DKS units, and was dedicated to operations with this Broshuis double-extending step-frame trailer, which was understood to be the first available in the north-east. The 40 foot trailer could be extended, by one man, to a staggering 74 ft 6 ins in 40-inch increments and was mainly used for oil pipe transportation. *(Photo: DAF Trucks Ltd)*

Despite the cross wind, this FTS3300 cracks on at a pace, performing the role of the express TIR machine it was designed to be. DAF's enormous roof spoiler and under-bumper air dam did much to cut fuel bills for those whose work involved a lot of motorway miles but, costing the better part of £500, they were not cheap and made up a significant part of the tractor's price if fitted.
(Photo: Marcus Lester)

The mighty edifice of the Space Cab looms large in this photograph taken at the Joppa transport café in Edinburgh. The owner-driven unit was a 6 x 2 and sported an eye-catching livery. In 1985, the Space Cab option added £2,750 to the price of a chassis plus £1,300 for the basic sleeper on which to mount it. The under-bumper air dam was a popular addition to many 3300s and came in at a far more reasonable £160, though the spotlights would be extra.
(Photo: John Henderson)

In a bold move in 1985, DAF took a convoy of eight trucks and buses on a promotional tour of Turkey and the Middle East. The three-month tour was designed to promote DAF's European truck design combined with the specific requirements of the region. Seventeen exhibitions were held en route between Istanbul and Alexandria. At the heart of the convoy was the 3300 in varying configurations and specs, including normal control and bonneted forms. The convoy obviously raised interest with a 250-truck order following for GH Transport of Dammam. The FT3300 Space Cab was the flagship of the convoy and is seen here undergoing final checks with other participants before departure and subsequently leading the group past intimidating pyramids. (Photos: DAF Trucks Ltd)

An EST driver grabs some well-earned rest while parked up behind the Edinburgh Playhouse during a Duran Duran concert in 1988. EST trucks have always been regular visitors across the border; such is the importance of Scottish venues on the UK tour circuit. B379 BPU was the third of EST's original batch of three FT3300 Space Cabs all of which completed around six years service with the company. Note the full-width roller blind, a unique feature to the Space Cab, which ran on guides parallel to the windscreen.
(Photo: John Henderson)

Eight new DAF cabs, including two destined for 3300 ATi chassis, leave the company's Westerlo cab plant in Belgium for the journey across the border to the Eindhoven assembly line. The Westerlo site was opened in 1966 and also housed the company's axle production plant. The other cabs here are the F220 for the medium-range trucks such as the FA2500 providing traction, and the F200 which was DAF's interpretation of the 'club of four cab', which was jointly developed with Volvo, Saviem and Magirus for use on lightweight distribution vehicles.
(Photo: DAF Trucks Ltd)

An impressive load, nicely roped and sheeted by the driver, is swept down the A30 courtesy of this smart 1984 FTS3300 of E Hesketh. While trucks and state of the art fighter aircraft seem poles apart, many of the big truck manufacturers, such as Volvo and especially Scania, enjoyed some involvement with the aviation industry. DAF was no exception. Its Special Products Division had been manufacturing landing gear for F-16 fighters since 1977 when, in 1984, it was awarded the contract to service the components of all 500 such aircraft operated by the USAF in Europe. This parallel business brought funds to the company and important aircraft technologies and practices could filter through to truck design and manufacturing.

(Photo: Marcus Lester)

Arthur Spriggs' FT3300 Space Cab looks slightly ungainly and certainly under-utilised pulling this single-axle box van trailer. Although this sort of job was usually entrusted to one of the middle-weight tractors on the Spriggs' fleet, operational requirements would sometimes find a premium tractor like the 3300 or a Scania 143 heading up one of these trailers for Smiths. Spriggs' Space Cab was amongst the first in the UK, and was notable for its short wheelbase which made the big cab look enormous. (Photo: Marcus Lester)

B650 FTF with a far more appropriate trailer. The unit's usual work was hauling PVC profile for use in double-glazing and plastic bottles for RPC Containers mixed with general haulage duties. In 1985 the unit, still in pristine condition, was borrowed by local DAF dealership, Peter Brown Ltd, to represent one of the new ATi tractors. During the show the unit wore the false number plate, D675 VNV and ATi badges. After the show, the unit reverted to its original registration but the ATi badges remained, making the unit something of a paradox. *(Photos: Arthur Spriggs & Sons Ltd)*

This fine FTT3326 was typical of a Middle East spec unit with high ground clearance and necessary extras such as the roof-mounted air-conditioning unit. DAF joined other high-quality foreign manufacturers such as Mercedes, Volvo and Scania in the Middle East market, where previously American machinery had been popular. This example was operated by BCI, which also ran 2800s. (Photo: DAF Trucks Ltd)

Far from competing with the airlines, Plane Trucking offered a complementary service used by such companies as TWA, KLM, Air France, Cathay Pacific and North West Orient to move airfreight from UK airports to the continent and vice versa. The usual load was made up of standard 96-inch air cargo pallets, but on occasions ran to more unique items such as a grand piano belonging to Barry Manilow and Formula 1 cars destined for the Grand Prix in Brazil. In 1985 the company ran a 100 per cent DAF fleet of eight FT3300 tractors and two drawbar combinations. The drawbars were highly specialised and were almost exclusively used to transport aero engines, two at a time, from the Rolls-Royce factory at Derby to Luxembourg. The tilt bodywork and Hydraroll roller bed floor eased loading and the entire combination was air suspended to provide the smoothest possible ride for the fragile and frighteningly expensive load. C548 EJX was the only Space Cab operated at the time. (Photos: Dave Wakefield)

DAF had been using its computer simulation system, TOPEC, since 1980 and it was invaluable in determining the performance requirement of the 3300 series of 1982. The system contained over 12,000 miles of information gathered from Western Europe's major routes. By telling the system the route, load, terrain and even driving style, an accurate prediction of fuel consumption and journey time for any given specification of DAF truck could be given. Whether the operator of this FTG3300 had envisioned hauling this remarkable trailer-mounted piece of plant, which looks as if it has come off the set of a 'Mad Max' movie, is not known, but it's doubtful that even TOPEC had a simulation for it. (Photo: Marcus Lester)

The cross members of the 6 x 4 chassis carried extra flitches of a saddle design to distribute loadings through the side rails and maximise flexibility. The 2699T double-drive bogie could be sprung with either parabolic or semi-elliptic springs, with the latter usually specified for off-road work. Despite the rigours of heavy haulage, this FTT3300 of A & I freight seems in fine condition. Note the fitment of non-standard super singles on the front axle. (Photo: Marcus Lester)

By the time this FT3300 was registered, the UK standard weight limit had increased to 38 tonnes and the 3300 started to make sense to many operators that had previously thought it just too much truck for their work. However, as the 3300 was at the forefront of the new breed of powerful yet economical trucks that manufacturers were designing in the early 1980s, it had always made sense, even at the old 32 tonne limit. Many who operated early examples found that with the eight-hour driving day, the big DAF could out-earn a lesser-powered truck. (Photo: Marcus Lester)

Hargrave International's slick operation depended on the performance of around fifty DAF trucks, all spotlessly turned out. MD Graham Eames was dedicated to providing the best service available and with other visionaries, such as Robin East, founder of Rokold, set up Transfigoroute UK for like-minded companies intent on improving standards and image in the fridge sector. Hardgrave naturally made the change to the 95 series in 1987, but also bought four FTG3300 tractors – surely the last available – five months after the new model's introduction. *(Photo: Adrian Cypher)*

The standard factory finish for the big Space Cab was the same gloss white as applied to all F241 cabs. This made for one big white cab, especially when fitted with the aerodynamic side screens. For those without a livery to apply, the mass of white could be broken up by a selection of factory decals offering a mixture of blue and red stripes. As well as the under-bumper air dam, this unit is also fitted with the side extensions, which ran under the first step. *(Photo: Marcus Lester)*

The ATi treatment was applied to DAF's heavy range in 1985 and for the 3300 that meant a new engine designation of DKX. Far from a mere marketing exercise, ATi was the latest reworking of the trusty old 11.6 litre engine which, through changes to breathing, injection, cooling and certain mechanical components, saw power rise further with even better economy than before. The maximum weight increase to 38 tonnes in 1983 brought about the biggest change to operating methods since the introduction of articulated vehicles in the 1950s and sparked a long-running debate in the industry as to how the new limit should be managed. For owner-drivers, it was often best to play safe with a three-axle unit that offered the flexibility to run with two- or three-axle trailers. DAF's twin-steer solution was no lightweight, especially when equipped with the Space Cab, but it was generally accepted that it was a high-quality solution, and the FTG chassis proved very popular with many examples working well into the late 1990s. (Photo: Adrian Cypher)

Different market conditions demanded many variations on the basic 3300 chassis. This outfit is fairly typical of a popular Dutch configuration, which provided the operator with a self-loading rigid tipper that doubled as the prime mover for his plant. The result was a combination that was self-supporting and could work away from base unassisted. The FAS3300 featured a 6 x 2 layout with a pusher axle and the tipper body could be swapped via hook loading. (Photo: Adrian Cypher)

Specialising in the international movement of delicate and expensive computer equipment allowed Allport Freight to operate a premium tractor like this FT3300 ATi, complete with Space Cab, at just 32 tonnes. To protect the precious cargo, both truck and trailer ran on air suspension. DAF's system was noted for its smooth riding characteristics, particularly on motorways, while retaining good anti-roll qualities. The optional set-up was available as a factory fitment direct from Eindhoven or as a UK modification. (Photo: Adrian Cypher)

In the days before double articulation by way of a dolly was legal, West of Scotland operated this system which used a fixed dolly with steerable wheels. In the absence of four-axle tractor units, this gave a standard 6 x 4 truck an extra capacity of 16 tonnes. If weight was not an issue, the dolly could be removed and the unit hitched directly to the fifth wheel as usual. The FTT3300 was found to be a great tool for this work, but the trailer, which featured a troublesome fluid-type suspension system, was far less reliable.
(Photo: Jimmy Campbell)

By the mid 1980s EST's three original FT3300 Space Cabs had performed so well that the fleet virtually standardised on the model until buying switched to the new 95 series post-1987. This meant fine, late model additions like this superb FT3300 ATi model, one of two bought together in 1986, with the new DKX engine and 354 bhp. As a typical load at the time was 12-14 tonnes, EST never felt the need to run the higher-powered 3600 model. (Photo: Adrian Cypher)

At a time when the automotive industry was enjoying great success with the GTi concept for cars, DAF marketing man Peter Symons saw the potential for applying something similar to trucks and came up with ATi. The ATi badges of the new model were prominently displayed on the grille and on the side of the cab. The launch was accompanied by a lavish advertising campaign to raise awareness of ATi in the industry. *(Photo: Marcus Lester)*

A nicely worn example of an FT3300 ATi operated by Lloyds of Ludlow. From 1985 on, the F241 cab benefited from major improvements at the Westerlo cab plant following a huge investment programme to ready the facility for the new Cabtec cab of the 95 series. The fully automated line took cabs from initial de-greasing to showroom gloss finish via a thirteen dip-tank process which applied, among other things, a zinc phosphate coat for even better corrosion resistance and involved three baking processes. The plant was then the most advanced in Europe and could produce a painted cab every seven and a half minutes – eighty per shift. *(Photo: Marcus Lester)*

Detrafor, like the Space cab, arrived in 1984 and derived its name from the owner's family name and its activities (Dens Transport en Forwarding). The Dens family was based in Antwerp and had been involved in shipping and transport since 1903. Detrafor was concerned with the movement of unaccompanied trailers shipped on RoRo ferrries, which were then tipped by UK-based trucks such as this smart FTG3300 ATi. Detrafor traded for nine years before the name was changed to Continental Cargo Carriers. (Photo: Marcus Lester)

DAF 3600
A True Flying Dutchman

In 1984 DAF produced its 250,000th truck. Significant indeed, but not as significant as the deal signed that year with Spanish, state-owned ENASA. ENASA, which controlled Pegaso (once Hispano-Suiza) had also been involved with International Harvester and quite recently had acquired one of the remaining British marques, Seddon Atkinson. The deal was for the joint development of a new cab system to suit the heavy- and medium-weight trucks of both parties.

Based at DAF's Eindhoven facility, CABTEC, as the venture was named, aimed to have the new product ready for the end of the 1980s. While this was great for the future, DAF's immediate problem was an ageing product and competitors that were constantly moving the goalposts of performance and driver comfort. The introduction of the second-generation cab in 1980 and the superb 3300 in 1982 had done much to revitalise the range, but another shot in the arm was going to be needed to keep DAF where they wanted to be for the next few years until the new designs came on stream.

The solution was a two-part programme, of which the first was the introduction of the Space Cab in 1985. The second part came back to engine development – always a DAF strong point – and was most significant as the results were really intended for the new models, but it was agreed that some would be released early to boost the existing range. The engine programme was called Ati (Advanced Turbo Intercooling) and its purpose was to provide powerful yet economical updates.

When the problem of more power was first raised in 1981, DAF's board of directors gave the R and D department at Eindhoven a simple remit: 'Keep pace with the competition, but outperform them in terms of fuel efficiency.' Luckily DAF had amassed a huge amount of data for all the major routes used by heavy trucks in Europe and started to run simulations on computers to represent long-distance journeys. When it assessed the data, it found that big-power trucks only used their full output for around 5 per cent of the time and that the relationship between power and journey times was hugely dependent on other factors outside the control of the driver and the power under his or her right foot.

Therefore, rather than take the easy option and just design a new, bigger engine with higher output, DAF's engineers would box clever and fine-tune the performance envelope of their existing unit to suit the real-world situations they had identified. Although the starting point was still the old faithful 1160 six-cylinder engine, this was to be the most comprehensive update yet, which started with a redesign of the cylinder block, crankshaft, piston rods and camshaft. The engine was given improved cooling for the block and pistons, and a further lowering of the temperature of intake air helped thermal efficiency, extending the life of components. Modifications were made to the inlet and exhaust manifolds which, combined with new turbochargers, gave better breathing characteristics. Finally, new injectors and modified valve and pump timings completed the picture.

The result was an overall reduction in engine speeds, lowering fuel consumption and extending reliability and service life, and a flattened, but extended torque curve which eliminated the peaky nature of the previous DKSE development. The 16-speed ZF gearbox became standard and a new axle, the single-reduction 1346 with low resistance properties, was introduced. As a result the new ATi trucks were more powerful and economical and because of the new torque characteristics, it was now easier to get the best from them too. In marketing terms ATi was a great success and created enough interest to sustain the range for another two years until the 95 series was ready.

The ATi range was launched at the Frankfurt Show in 1985 and comprised the DKZ Ati 1160, DKX Ati 1160 and the DKXE Ati 1160. The DKXE effectively replaced the old DKSE for the 2800 and the DKX was now the power unit for the 3300, but top of the pile was the DKZ, which, with a lower running speed of 2000 rpm, offered 373 bhp and 1047 lb/ft of torque with the flexibility to pull cleanly from as low as 900 rpm. It was this engine that was installed in the new flagship of the DAF range – the 3600. The non-ATi 2800 DKTD, with 253 bhp, was retained at the bottom of the range in the UK as an entry-level truck suitable for distribution-type roles.

The 3600 was no lightweight with a 4 x 2 unit, albeit with a generous 3.5 metre wheelbase and sleeper cab, weighing in at a shade under 7 tonnes; however, it did come well equipped. Heavy additions, such as a spare wheel and carrier were fitted as standard whereas many competitors conveniently left these items off when quoting chassis weight. Besides, the DAF chassis and cab were now renowned for quality and durability and most operators in the market for a premium truck accepted that this meant extra kilos. In any case, the weight was comparable with that of key competitors such as the Volvo F12 and Scania 142.

The ATi range received very favourable reports from the Press, smashing most economy records as it went, particularly the 3600, which proved that a well-designed, high-powered truck with a carefully mapped engine could be more economical than a less powerful one while providing excellent journey times. The 3600 found its own niche amongst operators, often becoming the crowning glory in fleets where both the 2800 and 3300 were also represented and with owner-drivers who wanted a premium truck with the best possible back-up behind it. Given the top-shelf nature of the 3600 and an all too brief production life, numbers on the road were not as great as that of other premium machines. To some extent the model's position was compromised when DAF and Leyland merged and brought together their respective ranges in 1987, but by then the big Dutch machine was about to move over anyway as the new 95 series was waiting in the wings.

The ATi range had its UK debut at the Scottish Motor Show in November 1985, one month after its launch in Frankfurt. With some of the toughest terrain and longest distances of any UK operators, it is little wonder that Scottish firms were among the first customers for the new FT3600 flagship. A typical example was G&J Jack, which early in 1986 added this example of the 373 bhp machine to its existing all-DAF fleet of five FT3300s. With the value of fresh fish loads dependent on how quickly they reached market, G&J Jack needed all the performance, reliability and back-up that it could get and found that DAF and its products consistently delivered for them. *(Photo: Adrian Lester)*

Another early example of an FT3600 operated by Woodbrow on bulk liquid work. The FT3600 came as standard with air suspension on the rear, the rest of the ATi range only having this as an extra which could be factory-fitted or carried out as a modification at DAF's extensive facility in Colchester. Opened in June 1983, DAF's £3 million pound Truck and Bus Centre in Essex handled all UK vehicle pre-delivery inspection, customer modification and delivery. The centre was also capable of down-plating products to suit specific operator requirements in the UK. *(Photo: Marcus Lester)*

C540 BFX started out as a demonstrator and was re-cabbed after being rolled into a ditch while on the continent. A new engine followed later so by the time Paul Hart took on the truck it was still virtually new. The replacement cab was unusual in that it was completely glazed, including the rear panel. The truck is pictured parked at Michaelwood Services on the M5 with a delightful Aquila step-frame tilt and tipping oil rig equipment from a similar trailer in Aberdeen. (Photos: Paul Hart)

Although closely resembling a factory Space Cab, closer inspection reveals that this FT3600 is actually sporting a replica of DAF's high roof extension. The give-aways in this case are the sharp edges of the design, its overall shape and the strip which runs around the middle, possibly added for strength or for joining the upper and lower portions. One notable after-market attempt at a Space Cab, which pre-dated the DAF item and possibly inspired it, was fitted to an FT3300 of H Kornherr transport of Vienna. (Photo: Adrian Lester)

What better way for DAF stalwart E Hesketh to celebrate fifty years in the haulage business than by running this fine FTS3600, DAF's ultimate development of the original 2800 which started the range twelve years earlier. Only by specifying the Space Cab option could this example have been trumped in the prestige stakes. Caught by camera on the A30 with one of Hesketh's customary high loads, D694 LJA makes an impressive three-plus-three combination. (Photo: Marcus Lester)

Initially the new 3600 was only offered as a 4 x 2 chassis in rigid (FA) or articulated (FT) form, which was fitted with the new 1346 single-reduction unit. Manufactured by DAF, the new unit had fewer moving parts, was quieter and, because fewer gears were engaged within, it offered less frictional losses. FTS (6 x 2 tag), FTT (6 x 4) and FTG (6 x 2 twin-steer) examples like this one were all added later. All early examples of six-wheeled chassis retained the old 2699 hub-reduction axle with only the very late examples benefiting from the 1346. *(Photo: Marcus Lester)*

By 1985, DAF's head of engine development, ED Sluyter, was in his 28th year with the company and nobody knew the 1160 engine better than he did. By identifying the six causes of loss that occur in an engine, Sluyter and his team were again able to squeeze more power from the existing engine. A fundamental part of the process was to lower the thermal loading in the cylinders. This was achieved by a number of changes that altered the cam profiles, inlet and outlet ports and valve timing. Combined with a lower running speed and improved scavaging and swirl characteristics in the cylinders, thermal loading was reduced, allowing more air to be pushed into the cylinders by a new design of turbocharger. This well-worked left-hand drive example is putting its 373 bhp to good use moving the mobile crane. *(Photo: Adrian Cypher)*

This FA3600 makes a fine combination with its drawbar trailer and impressive straw load. As a solo rigid the FA chassis was rated at 16.26 tonnes, but had the capacity to run up to 44 tonnes, where permitted, with a trailer. The flat-top nature of the chassis made the mounting of bodywork an easy affair. As rear cab windows were an option for the F241 by this time, those fitted to this unit would have added £140 to the price. Interestingly, the Space Cab was not offered as an option for the UK FA3600 chassis. *(Photo: Adrian Cypher)*

Two examples of McFarlane's FTS3600 units demonstrating the versatility of DAF's tag-lift offering to those on general haulage as they drop into Dover, one loaded and one not. The reduced drag of a raised axle when running empty could make a significant difference to fuel economy and tyre life. Early 3600s like these equipped with the old hub-reduction axle retained the 4.49:1 ratio; later units with the new 1346 unit benefited from a 3.31:1 ratio. *(Photos: Adrian Cypher)*

West of Scotland Excavations Ltd originally ran two low loaders to move its own plant and machinery to its own jobs, including the company's opencast mining operation, and to deliver machinery which was on hire from it. However, with the trucks drawing attention on the road more and more, outside jobs were coming in, resulting in the company's transport and plant hire departments fighting over which jobs were to be done first. The paying customer won out more often than not and West would do its own movements overnight. This smart FTS3600 was one of two such units bought to handle outside work. (Photo: Dave Wakefield)

West of Scotland's other FTS3600 unit, E879 KSX, travelling empty. As it also ran 2800 and 3300 units, West of Scotland completed the DAF hat-trick by operating examples of the 3600. With 60,000 sq ft of storage space at its base in Coatbridge, the company has ample room to tranship and store plant. It uses the services of two 40 tonne crawler cranes to assemble outsize loads. (Photo: Marcus Lester)

Somewhat under-challenged by this load, West's FTS3600 tractors would regularly operate up to the STGO Cat 2 limit. The company held the contract for moving all the production of the Terex plant at Newhouse and all the prime product of Caterpillar's factory at Tannochside. The Terex dumper trucks made especially impressive loads and reciprocal arrangements with both manufacturers kept their products on West's books for own-account use and hire.
(Photo: Jimmy Campbell)

Although the author doesn't know the operator of this fine FT3600, it is known that there was at least one other unit operated in the same smart livery, the consecutively registered E124 SAG. Both units featured the roof spoiler and side extensions which, mounted onto Space Cabs, made an impressive and fairly rare sight. This unit gained further wind-cheating properties by the addition of the under-bumper air dam, though the side extensions were not fitted. (Photo: Adrian Lester)

All ATi-powered trucks – not just the 3600 – featured a new intercooler which, with changes to tube diameter, the distance between them and their sectional shape, provided a smaller yet more efficient unit that reduced fan losses and improved the cooling capacity of the radiator. This, as well as allowing more power to be produced down the line by its effect, also reduced stress on the engine, helping to make the new ATi range as famously reliable as its predecessors. Cornering hard, this smart FT3600 was snapped on the dock road in Calais. *(Photo: Adrian Cypher)*

It was a bold move for DAF when, in 1979, they started to market the 2800 series trucks in Sweden. Despite the obvious strengths of the country's domestic product and a people with patriotic tendencies, the big DAFs sold well with the 3300 and the 3600 both following the wheel tracks of the 2800 into this difficult-to-crack market. Börje Jönsson Åkeri AB of Helsingborg on the south-west coast of Sweden was one operator that took on the Dutch machines. The company had begun in 1954 and grew rapidly through international haulage in the 1960s. Customer service and attention to detail are second to none and the company's trucks are always superbly turned out, as can be seen by this smart FT3600 heading back to the continent. *(Photo: Adrian Cypher)*

The remarkable success story of Eddie Stobart followed a switch to haulage from the family background in farming in the early 1970s. Significant growth was assured when a contract was won from Metal Box in the first year of operation. The first DAF truck was bought in 1974 and by 1985 the 40th Dutch machine, an FTG2800 DKSE, joined the fleet. E680 XAO was an impressive FT3600 with Space Cab, which, following long service, now enjoys a well-earned retirement basking in admirers who visit reception at Stobart's Daventry premises. *(Photo: John Henderson)*

Surely among the last 3600s to be registered in the UK, F137 TBD was an FA3600 operated by Rushden-based haulier Arthur Spriggs as a drawbar outfit. The 4 x 2 combination was new to Spriggs and originally employed on contract to Thurgar Bollé Ltd moving plastic products under that company's 'Thurbo' trademark and was sign-written accordingly. Later the truck was run in full Spriggs livery and was used to haul bubble wrap. The driver in this instance is Peter Spriggs, youngest of the five sons. *(Photo: Arthur Spriggs Ltd)*

The 3600 lived on in certain markets until 1993, long after the introduction of the replacement 95 series in Europe. DAF International Ltd, based in Manchester from 1989 until 1992, was set up to market vans, trucks and buses from the combined ranges of the merged Leyland DAF. Although mostly built to order, and generally kit-assembled locally, the 2800, 3300 and 3600 were all offered and proved popular in Asian, African and Far Eastern markets. To cope with rougher roads, springs were semi-elliptic and the cabs were strengthened, particularly the under-frame. Gearboxes offered were ZF 6-16 speed or Fuller 9-13 speed. The cabs featured a new 'three-bar grille' – which fitted the new corporate look – and a revised interior with a grey '95-style' finish. (Photo: DAF Trucks Ltd)

Although the export 2800, 3300 and 3600 were all badged as ATi, the 2800 was something of a hybrid featuring only some of the components of that development. The basis of that model was actually the old, original DKTD, which, in its low-stress state, proved extremely tolerant of low-grade fuel and oil. Many examples of the 2800 were operated in African states such as Kenya, Zaire and Zambia as articulated buses. The essential quality for the DAF export 3600 and its stable mates was the ability to be repaired and kept running with the most basic of facilities. Structural plastics, as found on the 95 series, were unacceptable, and the old F241 cab scored with its simple panels and small flat windscreen. (Photo: DAF Trucks Ltd)

Although the 3600 was no more as far as Europe was concerned after the introduction of the 95 series, the old models refused to die completely and back, by popular demand, in 1990 came two new developments to the old 1160. The 2900 DKV for Europe and the 3200 DKS for the UK featured the face-lifted cab of the export models with the 'three-bar grilles' and grey interiors. The new models joined an up-rated version of the old mid-weight 2500, the 2700 ATI, which, although powered by the 8.25 litre engine and fitted with the narrower F220 cab, effectively took the place of the old 2800. The 2900 and 3200 were never produced in any great numbers. Spence ran two of the latter on agricultural produce haulage from its base in Berwickshire. (Photo: John Henderson)

The 373 bhp 3600 ATI was the crowning glory of a range that started out in 1973 as the 248 and 307 bhp 2800. From that solid start, steady development, steered by careful analysis of the market and listening to operators and drivers, had created a superb range of trucks which had created a superb reputation for reliability and performance. DAF had come a remarkably long way in those years with trucks that held their own alongside the best of the best. As well as the competent products, much of the credit for DAF's rise has to go to its people, dedicated, enthusiastic and often young – if not in years, then in outlook. They were the ones who drove the company on. (Photo: DAF Trucks Ltd)

Other Books and DVDs from Old Pond Publishing

Scania at Work: LB110, 111, 140 & 141
Patrick W Dyer
This highly illustrated and informative book shows how Scania developed its models from the L75 to the 141, the truck that in the 1970s nearly every driver acknowledged to be King of the Road. Hardback

Know Your Trucks
Patrick W Dyer
'This is an excellent little book and does exactly what it sets out to do: that is, give enthusiasts the basics on which to begin to build their interest in trucks and the trucking industry.' *Trucking* magazine. Paperback

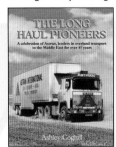

The Long Haul Pioneers: a celebration of Astran
Ashley Coghill
Tells the complete story of Astran but emphasises the early years when the drivers pioneered routes through to the Middle East. Over 350 photographs support a detailed and enthusiastic text. Hardback

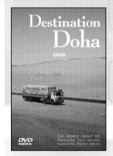

Destination Doha
'The World about Us'
In this 1977 programme meet the extraordinary men who formed the 5,000 mile transport link between Britain and the Arabian Gulf. Parts One and Two are both included in this two-disc set. DVD

Cola Cowboys
Franklyn Wood
The spirited account of what a UK journalist found when he accompanied truckers overland to the Middle East in the early 1980s. Paperback

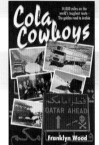

In the Driving Seat
Alex Heymer
Alex Heymer's driving experiences included the RAF, London Transport trolleybuses and buses, Galleon coaches and Gulf Oil tankers. Forty-four years driving rich in anecdotes. Paperback

www.oldpond.com
Old Pond Publishing Ltd, Dencora Business Centre, 36 White House Road, Ipswich IP1 5LT United Kingdom
Phone 01473 238200 • Full free catalogue available